Semiempirical Wave-Mechanical Calculations
on Polyatomic Molecules

A Current Review

A Chemistry–Physics Interface

Semiempirical Wave-Mechanical Calculations on Polyatomic Molecules

A Current Review

by Raymond Daudel and Camille Sandorfy

New Haven and London, Yale University Press, 1971

52896900

Designed by Sally Sullivan
and set in IBM Bold Face One type.
Printed in the United States of America by
The Carl Purington Rollins Printing-Office
of the Yale University Press.

Distributed in Great Britain, Europe, and Africa by
Yale University Press, Ltd., London; in Canada by
McGill-Queen's University Press, Montreal; in Mexico
by Centro Interamericano de Libros Académicos,
Mexico City; in Central and South America by Kaiman
& Polon, Inc., New York City; in Australasia by
Australia and New Zealand Book Co., Pty., Ltd.,
Artarmon, New South Wales; in India by UBS Publishers'
Distributors Pvt., Ltd., Delhi; in Japan by John
Weatherhill, Inc., Tokyo.

Contents

QD462
D381
CHEM

Figures

Tables

For a long time quantum chemists concerned with polyatomic molecules showed a rather marked preference for π-electron systems. This was due, apart from the well known, intriguing spectral and magnetic properties of conjugated molecules, to the relative simplicity of π-electron problems. The largely successful Hückel approximation makes it possible to regard the π electrons as living a somewhat independent life in the field of the nuclei. The other electrons may be relegated into the "core" and do not have to be considered explicitly.

This is indeed a very great simplification. In this approximation 1,3-butadiene is a problem with four π electrons while n-butane, with the same number of carbon atoms, is a problem with 26 valence electrons.

With the arrival of the computer era the situation changed radically. Since 1965 we have witnessed the dawn of what we may call the sigma- or valence-electron epoch in theoretical chemistry. It was our original intent, in early 1967, to write a review on σ-electron calculations, commenting on every relevant paper. (We knew of perhaps ten or twelve at that time.) The "publication explosion" in this field in 1968 and 1969 has forced us to renounce our original claim to complete coverage. We hope, however, that the most important stages in evolution are represented, up to the end of 1969.

The number of simultaneous discoveries by two or more authors has been unusual. This is another sign of the current general interest in σ-electron problems. We saw no reason to omit mention of the contributions of certain authors because of a few months of difference in publication dates.

In describing the various methods we have kept the original notations of the respective authors rather than making them uniform throughout the text. Ab initio calculations and diatomic molecules are mentioned only occasionally.

This book is meant to be complementary to Sigma Molecular Orbital Theory (SMOT) (O. Sinanoğlu and K. B. Wiberg), published by the Yale University Press in the Yale Series in the Sciences.

SMOT gives an extensive and overall view of the treatment of molecules with semiempirical as well as nonempirical methods and applications in organic and inorganic chemistry. Our present book is a detailed review of recent literature concentrating on semiempirical methods. Chapters 1 and 2 are reviews of the literature relating to Hückel and Pariser-Parr-Pople type calculations on σ-electron systems based on individual atomic orbitals. Chapter 3 is concerned with methods based on bond orbitals and polyelectronic functions. Chapter 4 is a brief register of recent developments contained in publications which appeared in 1969.

It is our pleasure to remember that our collaboration on this book began in 1968 when C. S. was a visiting professor at the Sorbonne attached to R. D.'s laboratory.

C. S. is indebted to Dr. D. R. Salahub for many helpful comments.

R. D. wishes to express his sincere thanks to Mme Jeannick Schuhé and C. S. to Mlle Gisèle des Groseillers and Mlle Hélène Brouillette for their excellent secretarial work.

May 1970

R. Daudel
C. Sandorfy

Abbreviations

AO	atomic orbital
CGTO	combination of Gaussian-type orbitals
CI	configuration interaction
CNDO	complete neglect of differential overlap
EH	extended Hückel
FSGO	floating spherical Gaussian orbitals
HF	Hartree-Fock
IA	ionization potential – electron affinity
IEH	iterative extended Hückel
INDO	intermediate neglect of differential overlap
LCAO	linear combination of atomic orbitals
LCBO	linear combination of bond orbitals
LMO	localized molecular orbital
MCZDO	many-center zero differential overlap
MINDO	modified intermediate neglect of differential overlap
MO	molecular orbital
NDDO	neglect of diatomic differential overlap
PDDO	projection of diatomic differential overlap
PNDDO	partial neglect of diatomic differential overlap
PPP	Pariser-Parr-Pople
RCNDO	Rydberg complete neglect of differential overlap
SCF	self-consistent field
SCGF	self-consistent group function
SLO	strictly localized orbitals
SMOT	<u>Sigma</u> <u>Molecular</u> <u>Orbital</u> <u>Theory</u>, by O. Sinanoğlu and K. B. Wiberg, Yale Series in the Sciences, Yale University Press, New Haven, 1970.
STO	Slater-type orbital
VESCF	variable electronegativity self-consistent field
WH	Wolfsberg-Helmholtz
ZDO	zero differential overlap
ZO	zero overlap

Chapter 1

Empirical Methods

The first applications of the simple Hückel molecular orbital
method to saturated hydrocarbons were made by Sandorfy and
Daudel (1), who used a method considering only the carbon skeleton (the "C" approximation), and by Sandorfy (2), who introduced all carbon sp^3 hybrids and hydrogen 1s orbitals (the "H"
approximation). We are going to recall the main features of
these methods and then follow subsequent developments. (At
this point the reader might like to consult a simple introducto-
ry review by Sokolov (3). See also Sigma Molecular Orbital
Theory (SMOT) for an overall view including nonempirical methods and chemical applications.)

1. The "C" approximation

The idea underlying this method is that, as Hückel was able to
extract the π electrons from conjugated molecules and treat
them as a separate problem, it may be possible to separate the
C–C bonds from the C–H bonds in saturated molecules and still
be able to account for some of the characteristic properties of
these molecules. There is clearly much less hope for such an
approximation to be successful than there was for the Hückel
method. While π orbitals have a nodal plane where σ electrons
have their greatest density, no such difference exists between σ
orbitals in C–C and in C–H bonds. Yet, one may hope that, in a
rough approximation, the C–H bonds in which the electrons are
more tightly bound than in the C–C bonds make a constant contribution to the total energy in different paraffin molecules and to
the electronic charge distribution in the C–C bonds.

Thus in the original "C" approximation, all H orbitals and
all carbon sp^3 hybrids linked to the H orbitals were disregarded.
For the remaining sp^3 orbitals, all the Hückel coulomb integrals
α_C were made equal, as were all the resonance integrals β_{C-C}
between atoms which are "chemically" bonded together. Non-
neighbor interactions and all overlap integrals were neglected.

1

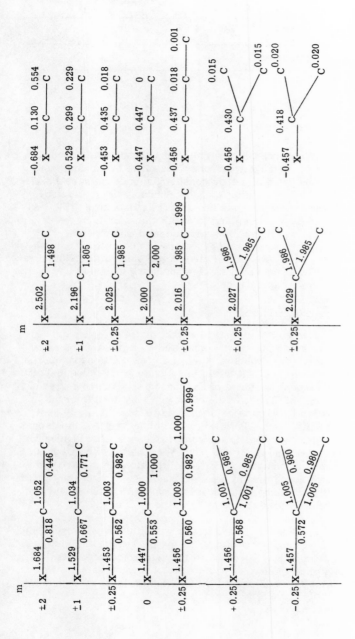

Fig. 1. Charge distributions using "C" approximation. Left to right: orbital charges, bond charges, net atom charges (Sandorfy (2)).

For the resonance integral between two sp^3 hybrids on the same carbon atom a parameter $\beta' = m\beta$ is needed. To obtain information as to the value of m, this latter parameter was varied. The " C" approximation gives orbital charges q_r all equal to unity for any value of m. If we substitute a heteroatom for one of the end carbon atoms and characterize it with a supplement to its coulomb integral ($\alpha_X = \alpha_C + \beta$ was used in obtaining the results given in Fig. 1), the result is that a heteroatom more electronegative than carbon attracts a large electronic charge from the carbon sp^3 orbital bonded to it but much less from more remote orbitals. In this case m = ± 0.25 gave the most reasonable values and the sign made a slight difference in the case of branched chains only.

Yoshizumi (4) determined the value of m empirically by fitting the calculated dipole moments of halogen-substituted derivatives of some normal paraffins to the experimental values. He found $m^2 = 0.12$ and was able to show that introducing the overlap integral between two "chemically" bonded σ orbitals does not change this value significantly.

Fukui, Kato, and Yonezawa (5) applied this simple approximation (with m = +0.34) with surprising success to the calculation of bond dissociation energies, total energies, and ionization potentials of normal and branched paraffins and many of their substituted derivatives.

They used $\alpha = \alpha_C - 0.3\beta$ for tertiary and quaternary carbon atoms (to fit the ionization potential of isobutane), set β proportional to the overlap integral between a heteroatom and a carbon atom, and determined the supplement to the coulomb integrals belonging to halogen atom X from the ionization potential relating to the C–X bonding orbital in methyl halides.

Taking the energy of the highest occupied orbital for the ionization potential, they found in the case of normal paraffins an almost perfect parallel with the experimental values. From a plot of the observed I_{pot} against the computed ones they derived for the two main Hückel parameters:

$$\alpha = -5.850\underline{ev} \text{ and } \beta = -6.364\underline{ev}$$

The fit was still good for branched-chain paraffins but, as expected, much less so for cycloparaffins containing small rings.

For heats of formation the use of the same parameters leads to a satisfactory parallel between the calculated and observed thermochemical values for normal and cyclic paraffins but less so for branched paraffins.

Using the "C" approximation, Fukui et al (6) studied nucleo-philic substitution reactions of alkyl halides (S_N1 and S_N2) and found good correlations between computed activation energies and observed rate constants.

Klopman (7–10) used a different set of parameters. He assumed that all C–H bonds are equivalent and that their energy is equal to one-fourth of the heat of formation of the methane molecule:

$$E_{C-H} = \frac{-392.857}{4} = -98.2142 \text{ kcal/mole}$$

For ethane, then, with its experimental heat of formation,

$$E_{C-C} = -667.017 - 6E_{C-H} = -77.732 \text{ kcal/mole}$$

On the other hand, since in the "C" approximation the secular determinant is simply

$$\begin{vmatrix} \alpha - E & \beta \\ \beta & \alpha - E \end{vmatrix} = 0 \qquad [1]$$

and $E = \alpha \pm \beta$ for one electron, and, for the ground state, the positive sign holds, we have that

$$2\beta = -77.732 \text{ kcal/mole}$$

and

$$\beta = -38.866 \text{ kcal/mole}$$

Thus the total C–C energy of a hydrocarbon can be calculated by subtracting the total C–H energy from the observed heat of formation.

We reproduce here a table contained in one of Klopman's papers (10) which shows the observed heats of formation of a few normal and branched paraffins (Table 1). The paraffins were formed from atomic carbon and hydrogen in the gaseous state at 0°K. In the last column heats of formation are expressed in units of $\beta = -38.866$ kcal/mole. It is seen that the heats of formation of paraffins are roughly but not exactly additive. This is one of the most important characteristics of paraffins, with which any theory concerning paraffins must agree. Klopman has shown that in order to match simultaneously both heats of formation and ionization potentials for all types of paraffins it is necessary to give a negative value, m = -0.36, to the parameter β' representing the interaction between two sp^3 orbitals on the same carbon atom. If we do this there is no need to use supplements to the coulomb integrals according to their being primary, secondary, tertiary, or quaternary.

Table 1. Observed heats of formation of normal and branched paraffins

	Energy of formation kcal/mole	Energy of C–H bonds kcal/mole	Experimental energy of carbon skeleton in units of $\beta = -38.866$ kcal/mole
Methane	−392.857	−392.857	0
Ethane	−677.017	−589.285	2
Propane	−943.612	−785.714	4.0626
Isobutane	−1223.06	−982.142	6.1987
Neopentane	−1502.69	−1178.570	8.3395
Cyclohexane	−1661.79	−1178.570	12.4330

After Klopman (10).

A second characteristic of saturated hydrocarbons, the pronounced decreasing trend of ionization potentials with increasing chain length, also serves as a test for all theories, and further verifies this approximation.

We may conclude at this stage that the "C" approximation, crude as it may be, correlates a surprising number of experimental characteristics of saturated hydrocarbons: heats of formation, bond dissociation energies, ionization potentials, and rate constants. It certainly should not be used indiscriminately, however.

The following point deserves mention. Klopman (9) has shown that, provided all coulomb integrals are the same, the introduction of a parameter $\beta' = m\beta$ between two hybrids on the same carbon is tantamount to introducing non-neighbor interaction 1,4 as shown in Figure 2. In fact it can be seen that

$$
\begin{vmatrix}
\alpha-E & \beta & 0 & 0 \\
\beta & \alpha-E & m\beta & 0 \\
0 & m\beta & \alpha-E & \beta \\
0 & 0 & \beta & \alpha-E
\end{vmatrix}
\equiv
\begin{vmatrix}
\alpha-E & \beta & 0 & m\beta \\
\beta & \alpha-E & 0 & 0 \\
0 & 0 & \alpha-E & \beta \\
m\beta & 0 & \beta & \alpha-E
\end{vmatrix}
\quad [2]
$$

Thus these are two equivalent ways of introducing a certain amount of delocalization or nonadditivity into the calculations.

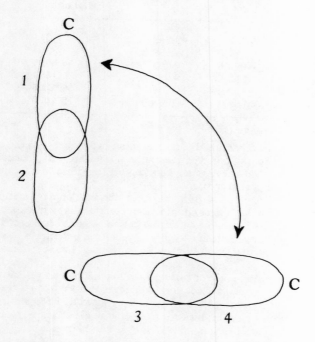

Fig. 2. Non-neighbor interactions between hybrid orbitals.

2. The "H" approximation

In the "H" approximation all valence electrons, that is, all carbon sp^3 orbitals and hydrogen $1s$ orbitals are taken into account. The parameters used by Sandorfy (2) in 1955 were $\beta_{C-C} = 1, \beta_{C-H} = 1.06$ proportional to the overlap integrals $S_{C-C} = 0.647$, $S_{C-H} = 0.684$; the coulomb integrals were taken equal to $\alpha_C = \alpha$ for all carbon atoms and to $\alpha_H = \alpha_C + (x_H - x_C)\beta$ where $x_C = 2.50$ and $x_H = 2.00$ are the Pauling electronegativities for carbon and hydrogen respectively. Overlap integrals between neighbors were retained so that the whole procedure resembled Wheland's approximation for π-electron calculations (11).

Fukui and his co-workers (6) applied a similar approach in which they put x_H equal to $\alpha_C - 0.2\beta$ to fit the bond dipole moment of the C–H bond, put β proportional to the overlap integrals, borrowed m = +0.34 from the "C" approximation and neglected overlap integrals and non-neighbor interactions.

With $\alpha = -5.05ev$ and $\beta = -8.8ev$ they obtained good agreement with observed ionization potentials and heats of formation, though somewhat less satisfactory than with the cruder "C" approximation. They used "frontier electron density" (the electronic charges in the highest occupied orbital) as a reactivity index and found good correlations with the speed of reactions of the type

$$R - H + X \rightarrow R\cdot + H - X$$

where X is an attacking radical (metathetical reactions).

It is interesting to study the charge distribution in individual molecular orbitals. Table 2 reproduces data on normal butane obtained by Fukui et al (6). The orbital $j = 13$ is the highest which is occupied in the ground state. In this orbital most of the electronic charge is accumulated in bonds 3, 4, and 7, that is, in the C–C bonds. This adds some weight to the supposition that when ionization occurs the electronic charge is taken essentially from the carbon skeleton (but see pp. 58, 100). The gap between highest occupied and lowest unoccupied orbital is about 9 or 10 ev with the presumed value of β, and it would be wide enough to place the whole set of π energy levels between them. This is in accordance with the usual assumption (which until recently was taken for granted) that the respective energies are in the order $\sigma \angle \pi \angle \pi^* \angle \sigma^*$. In the first empty orbital ($j = 14$) a significant amount of charge goes to the hydrogen atoms.

Table 2. Electron distribution in n-butane

```
        H   H H H
        |   | | |
   H ─ 2C³─4C⁷─C─C─H
        |   5| | |
        H   6  H H H
```

Orbital Energy j	ϵ_j	Electron distribution						
		$(C^j{}_{H_1})^2$	$(C^j{}_{C_2})^2$	$(C^j{}_{C_3})^2$	$(C^j{}_{C_4})^2$	$(C^j{}_{C_5})^2$	$(C^j{}_{H_6})^2$	$(C^j{}_{C_7})^2$
21–26	$\alpha-1.3722\beta$*	0.052	0.059	0	0	0.044	0.039	0
20	$\alpha-1.3612\beta$	0.005	0.006	0.047	0.048	0.074	0.066	0.092
19	$\alpha-1.3546\beta$	0.014	0.016	0.133	0.134	0.037	0.034	0
18	$\alpha-1.3436\beta$	0.012	0.013	0.104	0.101	0.005	0.005	0.200
17	$\alpha-1.0870\beta$	0.035	0.022	0.032	0.002	0.057	0.087	0.007
16	$\alpha-0.9473\beta$	0.070	0.032	0	0.032	0.012	0.027	0.083
15	$\alpha-0.7567\beta$	0.064	0.016	0.077	0.116	0.006	0.023	0.009
14	$\alpha-0.5822\beta$	0.026	0.003	0.099	0.058	0.008	0.063	0.114
13	$\alpha+0.6971\beta$	0.012	0.008	0.128	0.113	0.001	0.002	0.192
12	$\alpha+0.7339\beta$	0.016	0.011	0.151	0.155	0.024	0.033	0
11	$\alpha+0.7627\beta$	0.005	0.004	0.047	0.053	0.055	0.072	0.119
5–10	$\alpha+0.8322\beta$*	0.059	0.052	0	0	0.039	0.044	0
4	$\alpha+1.4135\beta$	0.014	0.031	0	0.037	0.107	0.050	0.014
3	$\alpha+1.5585\beta$	0.029	0.074	0.024	0.002	0.036	0.014	0.065
2	$\alpha+1.7073\beta$	0.023	0.070	0.094	0.064	0.022	0.007	0.004
1	$\alpha+1.8098\beta$	0.008	0.026	0.063	0.084	0.057	0.017	0.103

*Sixfold degenerate.
After Fukui, Kato, and Yonezawa (6).

In a later paper (12) Fukui et al. treated the σ skeleton of conjugated molecules in a similar manner. We reproduce the orbital, bond, and net atom charges they obtained for 1,3-butadiene (Fig. 3). It is interesting to note that atom charges around the hydrogens are smaller in benzene and naphthalene than in ethylene or butadiene, and are less positive for the hydrogens of the methyl group in toluene. This simple method is capable of interpreting surprisingly fine details, as is sometimes the π-electron Hückel method.

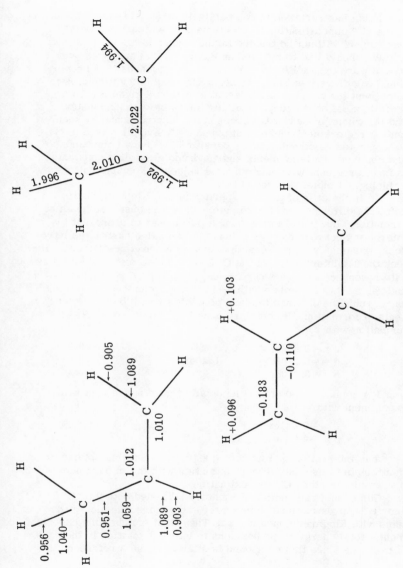

Fig. 3. Charge distributions for 1,3-butadiene: σ-orbital charges, σ-bond charges (sp² hybrids), and net atom charges (Fukui et al. (12)).

Fukui has reviewed in two publications (13, 14) applications of the "H" approximation to problems of chemical reactivity; this subject will not be treated further here.

We should like to mention an earlier (1958) and often over-looked work by Del Re (15). He put the differences in α proportional to differences in electronegativity, and in addition took into account the effect of all adjacent atoms on a given α. This leads to a set of interdependent auxiliary coulomb parameters, and the whole procedure approaches the electronegativity equalization requirement which is at present often applied (see p. 39). He computed electronic charge densities for many compounds and obtained dipole moments and quadrupole coupling constants in fair agreement with experimental values. (For applications to biological molecules see (16).)

As in the case of the "C" approximation, Klopman (8, 10) defined a different set of parameters based on observed heats of formation. Instead of introducing a parameter to represent the interaction between two $\underline{sp}^3$ hybrids on the same carbon, he introduced parameters for interactions between nonadjacent atoms. He took the 1,4 interactions of the C–C–C, C–C–H and H–C–H types with resonance integrals $c\beta_{C-C}$, $d\beta_{C-C}$ and $p\beta_{C-C}$ respectively and varied these together with $\beta_{C-H} = t\beta_{C-C}$ and $\alpha_H = \alpha_C + n\beta_{C-C}$ until he obtained the best possible fit with the observed heats of formation. He obtained excellent results for heats of formation with

$$c = -0.3305 \qquad t = 1.2375$$
$$d = -0.1101 \qquad n = 0.4049$$
$$p = -0.0806$$

For ionization potentials he obtained good agreement with experiment with

$$\alpha = -7.641\underline{ev}$$
$$\beta = -4.226\underline{ev}$$

It is interesting that no such good agreement was obtained in attempts to use the "H" approximation with the m parameter borrowed from the "C" approximation.

Cambron-Brüderlein and Sandorfy (17) used both the "C" and "H" approximations to interpret some stereochemical problems with Klopman's parameters. They first matched the experimental total energies and barriers to internal rotation in the case of n-butane since this compound is often taken as a reference in conformational problems. They did this by varying slightly the coulomb integral α according to the non-neighbor interactions by which a given orbital is affected. This is much easier than introducing small β's between non-adjacent orbitals and it can be justified by reference to the terms representing interactions between an electron on a given core and the other cores which are contained in α. They applied the parameters determined in this way to a number of saturated hydrocarbons, even steroids, and obtained good agreement with available experimental data.

3. The extended Hückel (EH) method

Many years ago Mulliken (18) and Wolfsberg and Helmholtz (19) suggested a very simple type of parametrization for Hückel calculations which greatly facilitates their extension to σ-electron problems. The coulomb integrals for an orbital i ($\alpha_i \equiv H_{ii}$) are taken as the appropriate valence-state ionization potentials and the resonance integrals ($\beta_{ij} \equiv H_{ij}$) are expressed as

$$H_{ij} = 0.5 \, K \, (H_{ii} + H_{jj}) \, S_{ij} \qquad [3]$$

All that is needed for this are overlap integrals, which can always be calculated, and valence-state ionization potentials, which most workers take from the tables of Pilcher and Skinner (20) or Hinze and Jaffe (21).

Hoffmann (22) took up this method and with a very rapid computer program he was able to apply it to a wealth of molecules with all valence electrons taken into account. Instead of using hybrid orbitals, as had previous authors, he used pure atomic orbitals and included all interactions and all overlap integrals. The constant K was chosen to be 1.75 in order to give the best agreement with the greatest possible number of experimental values. This is a crude method but it has extreme versatility. It can be applied without difficulty to three-dimensional molecules with the possibility of varying the assumed geometry. It soon revealed itself as an effective method of exploring equilibrium geometries, potential surfaces, energy differences between conformers, and other properties of three-dimensional molecules. As Hoffmann put it: "A Hückel calculation of this type, carried out as a function of inter-nuclear distance, gives rise to a potential curve having a minimum not far from the correct experimentally determined geometry of the molecule." Binding energies are in less satisfactory agreement with experimental values and electron charge densities are usually too high.

EH calculations of molecular geometries can play the role of Walsh's rules (23) for polyatomic molecules, an invaluable help to spectroscopists. These calculations are easily amenable to further refinements.

In subsequent papers Hoffmann discussed σ orbitals in azines (24), suggesting that the lone-pair concept is inadequate and that there is a considerable amount of delocalization in the highest occupied σ orbital; he also discussed compounds of boron and nitrogen (25) and carbonium ions (26). Hoffmann, Imamura, and Hehre (27) explored interactions between radical lobes in the same molecule separated by a number of σ bonds for benzynes and dehydroconjugated molecules. Adam and Grimison applied the EH method to the problem of nucleophilic substitutions in pyridine, quinoline, and isoquinoline (28) and the electrophilic substitution in imidazole (29); Adam and co-workers to hetaryne intermediates (30). Recently, Paldus and Hrabé (31) used it to determine the force field for the vibrational problem of ethylene.

Jordan and Pullman (32) applied the EH method to a study of the conformation of nucleosides of the purine and pyrimidine bases of nucleic acids and found the preferred conformations in agreement with available experimental data.

Hopkinson et al. (33) used it for the interpretation of reactions involving charged species (protonation reactions of acids, amides, and esters). Murthy et al. (34) applied it to the study of hydrogen bonding in methanol and formic acid. (They also used the CNDO/2 method for this; see p. 51.)

In order to justify the extended Hückel theory by comparison with ab initio SCF MO wavefunctions and one-electron energies, Allen and Russell (35) carried out calculations on BeH_2, BH_2^+, BH_2^-, NH_2^+, H_2O, BeH_3^-, BH_3, CH_3^+, Li_2O, F_2O, $LiOH$, and FOH. They used the EH theory with

$$H_{ij} = KS_{ij} (H_{ii}H_{jj})^{1/2}$$

which is one of Mulliken's suggested formulas (18) later used by Ballhausen and Gray (36). (Allen and Russel (35) give many useful references for works of Allen and his collaborators and for other papers dealing with EH calculations.)

As Allen and Russell put it: "Fundamentally, Hückel theory is an attempt to reproduce, by means of a one-electron model, the one-electron results of the many-electron Hartree-Fock solution." The Hartree-Fock (HF) method yields accurate equilibrium bond angles (within 1 or 2%) and makes good predictions for mo-

lecular shape. Allen and Russell show that this is because bond angles are determined by a simpler quantity than the total energy if one of the following inequalities is satisfied:

$$\frac{\partial}{\partial \theta} \left(\sum_i \epsilon_i \right) \Big/ \frac{\partial}{\partial \theta} \left(V_{NN} - V_{ee} + \sum_i \epsilon_i \right) > 0 \qquad [4]$$

or

$$\left| \frac{\partial}{\partial \theta} \sum_i \epsilon_i \right| > \left| \frac{\partial}{\partial \theta} \left(V_{NN} - V_{ee} + \sum_i \epsilon_i \right) \right| \qquad [5]$$

where V_{NN} represents the nuclear repulsions, V_{ee} the two-electron coulomb and exchange potentials, and the summation is for all valence orbitals. If the sign of the derivative with respect to the bond angle θ in the region of the energy minimum is the same as that of the total energy, $\sum_i \epsilon_i$ yields the same angle as the total energy. Except for highly ionic species and for certain cases of double minimum potentials, the inequalities hold in the HF approximation. Herein lies the chance of one-electron theories to make correct predictions of bond angles and molecular shape.

Calculations on individual molecules show that the relation between angles and one-electron MO energies depends essentially on molecular symmetry type and lead to a justification and extension of Walsh's rules (23). There is a far-reaching similarity between HF and EH results as to the order of energy levels, molecular shape, and changes in hybridization and ionization potentials with angle. The simple inequality relations do not apply to other molecular properties (the ionization potentials themselves, bond lengths, dipole moments, quadrupole moments, excitation energies) and Hückel calculation may lead to serious errors. Allen and Russell conclude that improvements could be achieved by using a HF atomic orbital basis set, making the charges self-consistent and correcting the one-electron energies in the case of polar bonds by a term "proportional to the electronegativity difference of the bonded atoms divided by the interatomic separation and summed over the bonds."

4. The iterative extended Hückel method

The logical next step seems to be to make the EH method self-consistent as to charges; that is, to make it iterative in the same sense as was done for a long time in π-electron calculations (37–40). This has been done by several authors (41–43). A typical approach is the one of Rein et al. (44). They supposed that the coulomb integral α has a linear dependence on the net atomic charge q:

$$\alpha = \alpha_0 + q\Delta\alpha \qquad [6]$$

and they obtained α values at each iteration from those of the previous cycle by the equation

$$\alpha_{n+1} = (1-\lambda)\alpha_n + \lambda(\alpha_0 + q_n\Delta\alpha) \qquad [7]$$

where λ is the damping parameter taken to be about 0.1.

They computed β by Cusachs' formula (see below) from the α at each iteration, and the overlap integrals from Slater atomic orbitals with exponents computed by Clementi and Raimondi (45). The latter were computed from SCF functions with a minimum basis set and take account of the screening by the electrons "outside" of a given electron.

L. C. and B. B. Cusachs (46) made a successful attempt to introduce into one-electron theories the ideas of Ruedenberg's analysis of chemical binding (47, 48). They obtained molecular orbitals as usual as linear combinations of atomic orbitals; minimization of the energy with respect to the LCAO coefficients leads to

$$\left| H_{ij} - ES_{ij} \right| = 0 \qquad [8]$$

Overlap integrals are retained and great care is given to the choice of atomic orbitals which are used to compute them. The best available atomic orbitals are the many-term SCF functions with only one orbital exponent for a limited range of interatomic distances (49). Another possibility is to choose the orbital exponent to match some atomic property. However, the values of many one- and two-center quantities depend strongly on the choice of Z and it is preferable to use different orbital exponents for molecular overlap and for atom-like terms depending on the shape of the orbital near the nucleus. Examples are given by Ruedenberg and Edmiston (48).

Valence-state ionization potentials are used to determine the coulomb terms, the H_{ii}. They need rescaling when the atom is in the molecule (see the discussion of Brown et al. in section 1 of Chapter 2), and the kinetic and potential parts should be treated differently. It is usually sufficient, however, to adjust the H_{ii} for the final charge which can be anticipated even without iterations.

The most important innovation of Cusachs concerns the off-diagonal matrix elements H_{ij}. Slater (50) has shown that in the case of the hydrogen molecule, what prevents the molecule from contracting a great deal beyond the equilibrium separation is the rapid increase in the kinetic energy of the electron rather than electron-electron or nucleus-nucleus repulsion. On the other hand, Ruedenberg (51) has shown that the two-center kinetic integrals, T_{ij}, are approximately proportional to the square of the overlap integral rather than to its first power.

Now, one can suppose (46) the virial theorem is approximately valid for the diagonal terms, H_{ii}, computed from atomic orbitals in a screened coulomb field:

$$H_{ii} = T_{ii} + V_{ii}$$
$$T_{ii} = -V_{ii}/2$$
$$H_{ii} = V_{ii}/2$$

Then for the potential part of the Mulliken-Wolfsberg-Helmholtz type H_{ij} we can write that

$$V_{ij} = S_{ij} (V_{ii} + V_{jj})/2 = 2 S_{ij} (H_{ii} + H_{jj})/2 \qquad [9]$$

For the kinetic part, Ruedenberg (51) found that the following formula applied in very good approximation in the cases he examined:

$$T_{ij} = S_{ij}^2 (T_{ii} + T_{jj})/2 \cong -S_{ij}^2 (H_{ii} + H_{jj})/2 \qquad [10]$$

Combing the two, one obtains

$$H_{ij} = S_{ij}(2- \left| S_{ij} \right|) (H_{ii} + H_{jj})/2 \qquad [11]$$

This is the formula used by Cusachs for the off-diagonal matrix elements in EH calculations and it is probably a major improvement, taking into account the different behavior of the kinetic and potential energies with the variation of interatomic distance. A local relative coordinate system has to be used in order to achieve invariance under rotations of the molecular coordinate system.

The reader will probably want to follow more closely the interesting arguments to be found in the Cusachs' paper (46) as well as in earlier publications on H_2O and NH_3 (52), and on the selection of molecular matrix elements from atomic data (43).

Carroll, Armstrong and McGlynn (53), working on H_2O and H_2S, varied the orbital exponent in calculating the overlap integrals after every iteration to take account of the atomic charges. The coulomb integrals too were corrected for these. Their results agreed well with those of ab initio calculations and led them to a reasonable interpretation of the electronic spectra of H_2O and H_2S.

In another paper Cusachs (54) showed how upper and lower bounds can be found for the absolute magnitude of H_{ij} and examined current molecular orbital theories in the light of these. A lower bound results from the requirement that the order of energies follows the number of nodes, that is, antibonding orbitals should have energies higher than the bonding ones. An upper bound can be given since, in a homopolar diatomic bond, the antibonding orbital must be below the ionization potential and, in a partially ionic bond, between the energies of the atomic orbital of higher energy and the ionization limit.

Duke (55) proposed a slightly different version of the iterative EH method and applied it to BH_4^-, NH_4^+, CH_4, C_2H_6, B_2H_6, C_2H_4, and C_6H_6. He compared the results to those obtained by the noniterative method and to published HF results and found that iterations do not always improve the results.

Pollak and Rein (56) noticed in some π-electron calculations that the results of iterative calculations are different according to whether Mulliken's (57) or Löwdin's (58) definition of the atomic charge is adopted. Cusachs and Politzer (59) were led to similar conclusions in their all-valence electron calculations on diborane. The differences became dramatic when hydrogen $2p$ orbitals were added to the bases set. Later (60) they examined the problem by studying the actual electronic distributions corresponding to the total wave functions in the manner of Roux et al. (61). Hillier and Wyatt (62) examined different ways of partitioning the overlap charge density in molecular orbital calculations.

Pullman and his co-workers (63) made an extensive study of purines and pyrimidines using iterative EH calculations. A comparative study of different all-valence electron calculations on biologically important purines and pyrimidines was reported by Pullman (64).

An interesting Hückel study was made on DNA by Ladik and Biczó (65). Fischer-Hjalmars et al. carried out semiempirical studies on metal complexes of biological importance (66). Adam and Grimison (67) examined σ polarization in five-membered heterocyclic ring systems.

Many other references can be found in the papers mentioned above.

There are a number of papers attempting to apply the EH method to coordination complexes (68–73, 41). Jörgensen and his co-workers (74) examined the influence of interatomic coulomb interactions (Madelung interactions) on Wolfsberg-Helmholtz calculations. Fenske, Radtke, and their co-workers (75, 76) presented an improved semiempirical SCF MO method for closed-shell systems. These works lie outside of the scope of the present review. They constitute, however, a closely related and very important field.

5. <u>Simulated nonempirical LCAO MO SCF methods</u>

Newton, Boer, and Lipscomb (77) worked out a nonempirical procedure in which the integrals are estimated rather than computed theoretically; no experimental data are used to adjust the estimates. The integrals are chosen so as to match the results of the exact Roothaan LCAO MO SCF method (for closed shells, with a minimum basis set and Slater orbitals) for simple molecules as closely as possible, and then these are used in calculations on larger molecules. This idea of "simulating" the Hartree-Fock-Roothaan method also appears in the CNDO scheme described in Chapter 2, but it does not play such an essential role.

The α's for larger molecules are taken from exact calculations for small molecules in which the given atom has the same environment as much as possible. These α's are then used for computing the β's from the formula ($\alpha_i \equiv M_{ii}$, etc.)

$$M_{ij} = K_{ij} S_{ij} (M_{ii} + M_{jj})/2 \qquad [12]$$

A major point in this method is the realization that the different parts of M_{ij} require different K_{ij} values (see also Cusachs (46). This is particularly important in the case of the kinetic energy, as can be seen from Table 3 relating to methane (77). It is based on the exact SCF calculations of Palke and Lipscomb (78). In kinetic energy terms the variation of K (and therefore, the total Hamiltonian) is so great that β cannot possibly be calculated with just one or even two K's ($K_{\sigma\sigma}$ and $K_{\pi\pi}$). On the other hand, for

Table 3. K_{ij} for two-center matrix elements of methane

	Total Hamiltonian	Kinetic energy	Potential energy	Nuclear attraction energy	2-Electron interaction energy
K_{1sH}	2.04	−0.01	0.83	0.92	1.14
K_{2sH}	1.46	0.37	1.05	1.00	0.98
K_{2pH}	2.10	0.54	1.00	0.93	0.91
K_{HH}	2.95	−0.04	1.19	1.09	1.05

After Lipscomb et al (77).

the potential energy terms the K are reasonably constant. Thus Newton et al. computed the kinetic energy parts of the α's (and the overlap integrals) exactly and used them to obtain the β's through equation [12]. Table 3 lists K values for the potential energy terms based on available exact calculations. Most of these are fairly constant from molecule to molecule, but distinct values are needed for $K_{2p\sigma, 2p\sigma}$ and $K_{2p\pi, 2p\pi}$ for the π systems of planar molecules.

One-center two-orbital elements are given separate consideration. The 2s-2p elements have values of several electron volts and are needed to simulate SCF calculations accurately. These are zero overlap (ZO) elements and are computed from the formula $F_{aa} = K_{aa}^{ZO} \sum_k S_{ak} S_{a'k} \alpha_k$ where a and a' are two (orthogonal) functions on center a, S_{ak} and $S_{a'k}$ are their respective overlap integrals with all the basis functions, and K is taken as 0.5 for the 2s-2p element. $F_{1s, 2s}$ is treated as a two-center element, since the overlap integral is not zero between 1s and 2s (Slater orbitals). $K_{1s, 2s}$ turns out to be 0.66. All electrons are included. Then the Roothaan SCF scheme is used to compute eigenvalues and eigenvectors.

This method has many interesting characteristics. No empirical data are used to fix the parameters. This implies freedom from many of the arguable approximations which are generally applied, such as the ZDO approximation, the use of valence-state ionization potentials for obtaining the α, the Mulliken approximation for integrals containing differential overlap, and the assumption that the α's vary linearly with net atomic charge and that the β's are proportional to overlap integrals. There is a good discussion of all these characteristics in Newton, Boer, and Lipscomb (77).

On the other hand this method cannot be better than the complete Roothaan LCAO MO SCF method, and does not take into account the correlation by empirical adjustment of certain integrals.

The invariance requirement with respect to the choice of local axes for the 2$\underline{p}$ atomic orbitals was met by two approximations: (1) that an average value of α be used for all three 2$\underline{p}$ orbitals on a given atom and, (2) that the coefficients K_{ij} governing $\underline{s},\underline{p}$ and $\underline{p},\underline{p}$ interactions be the same for all three 2$\overline{\underline{p}}$ orbitals on each atom.

Boer et al. applied their method to boron hydrides (79), hydrocarbons, and many carbonyl and heterocyclic compounds (77). They computed dissociation energies, ionization potentials, charge distributions (both π and σ), conformational and isomerization energies, and dipole moments, and examined the concepts of lone pair, conjugation, hyperconjugation, and charge transfer. They used two approximate criteria for the validity of their results, the molecular binding energy and the virial theorem. The latter is made possible by their separate handling of kinetic energies.

Among their many interesting results we mention that the lowest π orbital in aromatic molecules has a lower energy than several σ orbitals. (Hoffmann, who used the extended Hückel method, was led to the same conclusion (22).) This leads to a new interpretation of the Rydberg states of benzene and pyridine. Another interesting finding is the appreciable delocalization of the lone pairs in carbonyls and heterocycles. (See Brown and Harcourt (80).)

For a detailed discussion of charge distributions we refer the reader to Newton's article (77).

6. The causes of σ-electron delocalization

In a recent paper Pople and Santry (81) presented a critical Hückel study related to saturated molecules. Their basic atomic wavefunctions are the hydrogen 1$\underline{s}$ functions and the carbon 2$\underline{s}$ and 2$\underline{p}$ functions which (in their first method) they do not hybridize. They assume that the diagonal matrix elements (the α) depend on the nature of the atomic orbitals only (whether

H1$\underline{s}$ or C2$\underline{p}$ or C2$\underline{s}$); so there are only three parameters of this kind $\alpha_{\underline{h}}$, $\alpha_{\underline{s}}$, and $\alpha_{\underline{p}}$. As to the nondiagonal matrix elements (the β), they assume that they are equal to zero if the two atomic orbitals are at the same carbon. This is reasonable if we think of the resonance integrals as proportional to the overlap integrals since atomic orbitals belonging to the same atom are mutually orthogonal. C–H bonds introduce $\beta_{\underline{sh}}$ and $\beta_{\sigma\underline{h}}$, where σ stands for a 2$\underline{p}\sigma$ orbital having its axis directly along the C–H bond. If the axis of a $\underline{p}$ orbital makes an angle θ with the C–H bond, the integral is $(\cos\theta)\beta_{\sigma\underline{h}}$. The C–C bonds introduce four other integrals, $\beta_{\underline{ss}}$, $\beta_{\sigma\underline{s}}$, $\beta_{\sigma\overline{\sigma}}$, and $\beta_{\pi\pi}$ where the axis of the π function is perpendicular to the bond. If two functions are inclined at angles θ and θ' to the bond axis and have an azimuthal angle difference of ϕ the integrals are:

$$\beta_{\underline{pp}} = (\cos\theta\ \cos\theta')\beta_{\sigma\sigma} + (\sin\theta\ \sin\theta'\ \cos\phi)\beta_{\pi\pi'} \qquad [13]$$

All integrals are normally negative except for $\beta_{\sigma\sigma}$.

Using hybridized orbitals instead of the carbon 2$\underline{s}$ and 2$\underline{p}$ functions leads to exactly the same results since these functions are linked together by an orthogonal transformation. Figure 4 shows the necessary integrals involving tetrahedral hybrids as given by Pople and Santry. These authors then perform a normal Hückel calculation computing electronic charge densities and bond orders. The existence of long-range bond orders is shown to be "an indication of the intrinsic delocalization of the bonding electrons."

Pople and Santry carried out two calculations. In the first one they set

$$\alpha_{\underline{s}} = \alpha_{\underline{p}} = \alpha_{\underline{h}} = 0$$

This implies that 2$\underline{s}$ and 2$\underline{p}$ atomic orbitals have the same energy so that complete hybridization may take place, and that there is no polarity in the C–H bonds. Furthermore they neglect all interactions between orbitals on atoms which are not "chemically" bonded together. Under these circumstances normal paraffins and cycloparaffins without odd-numbered rings behave like alternant molecules, and many of the theorems valid for alternant π-electron systems apply to them. Thus the molecular orbital energies occur in equal and opposite pairs and charge densities are equal to unity.

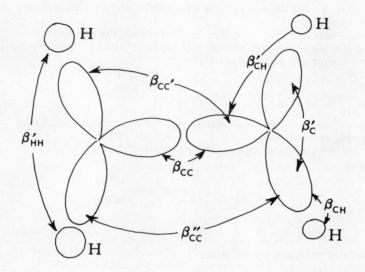

Fig. 4. Tetrahedral hybrid orbitals between which matrix elements are required (Pople and Santry (81)).

With these approximations we obtain completely localized bonds, as is immediately seen if we use the tetrahedral hybrids; the secular determinant factors into 2×2 units, one for each bond. This would lead to complete additivity for the heats of formation and to the same ionization potential for all paraffins. Also the nuclear spins of non-neighbor atoms would not influence one another. All these facts are known to be contrary to experiment.

So Pople and Santry went on to reintroduce the elements causing delocalization in the form of a perturbation calculation. They took for the diagonal elements

$$\alpha_s = 16.0\underline{ev}$$
$$\alpha_p = 11.2\underline{ev}$$
$$\alpha_h = 13.6\underline{ev}$$

these values corresponding to the respective ionization potentials in the separate atoms.

The overlap integrals in terms of hybrid Slater orbitals are (Fig. 4):

$$S_{CC} = 0.647 \qquad\qquad S_{CH} = 0.688$$
$$S'_{CC} = 0.108 \qquad\qquad S'_{CH} = 0.150$$
$$S''_{CC} = 0.007 + 0.128\cos\phi$$

Nonbonded overlaps are quite appreciable. The last one introduces an element $\beta_{\pi\pi}$ with some π-bonding contribution in C–C single bonds. Non-neighbor β's may also be appreciable.

The perturbation calculation disclosed that delocalization arises from three causes: (1) the difference between the energies of atomic 2s and 2p orbitals, (2) bonding between non-neighbor atoms, and $(\overline{3})$ partial π bonding in the C–C bonds. The two first effects cause geminal delocalization but they tend to cancel each other. The third causes vicinal delocalization which is greatest for trans configurations. The total change in energy due to delocalization is quite appreciable, 64 kcal/mole in the case of n-butane, for example. This is greater than the 37 kcal/mole which would be due to dispersion forces between nonbonded atoms, according to Pitzer and Catalano (82). Pople and Santry were able to show that, although the delocalization correction on the energy is appreciable, this is consistent with the approximate additivity of bond energies.

Chapter 2

Semiempirical Methods

The logical next step in the study of σ-electron systems is to adapt the Pariser-Parr-Pople method for the treatment of such systems. This method, highly successful with π-electron problems, involves neglect of overlap and differential overlap and we ask if the approximations are justified in the case of σ electrons or if we have to move closer to Roothaan's full LCAO SCF method, or the antisymmetrized molecular orbital and configuration interaction method, or a combination of them.

The neglect of overlap integrals in the π-electron PPP method is justified by assuming that our starting atomic orbitals were actually orthogonalized Wannier-Löwdin orbitals and the neglect of differential overlap by cancellations between certain electron-repulsion integrals caused by the intervention of the overlap integrals (which appear in the Wannier-Lowdin orbitals). In the case of σ bonds, some of the overlap integrals have very high values. We still may find justification for the use of zero-differential overlap methods. Other possibilities are also open to us. Mulliken-type approximations and atomic spectroscopic data can be used. The choice of the basis set may change certain aspects of these methods.

In recent years a number of attempts have been made to apply semiempirical methods to saturated molecules. We will describe these in some detail.

1. The VESCF method

Brown and Heffernan (83–85) pointed out in 1958 that when the electronic charge density changes on an atom the core coulomb integral α is not the only one to change. The charge has an influence on the orbital exponent itself and therefore all selected inte-

---

Note: For some of the key papers mentioned here and related theory, the reader might find it useful to consult Sigma Molecular Orbital Theory.

grals must change. This idea is the basis of their VESCF (variable electronegativity SCF) method which they first used in π-electron problems. They regarded Z_μ as a function of q_μ and they used Slater's rules for establishing a relation between them:

$$Z_\mu = N_\mu - 1.35 - 0.35(\sigma_\mu + q_\mu) \qquad [1]$$

for first row elements. The underlying assumption is that Slater's rules apply for continuous variation in electron density. N_μ is the atomic number of atom μ and σ_μ is the number of σ electrons contributed by μ. Through Z_μ, integrals like $(\mu\mu|\mu\mu)$, $(\mu\mu|\nu\nu)$, and $(\nu:\mu\mu)$ become functions of q_μ and so does the valence-state ionization potential, I_μ.

In their calculations Brown and Heffernan neglected differential overlap, computed two-center coulomb integrals from a Pariser-Parr type extrapolation formula (86–88), and computed the one-center coulomb integral from Paoloni's formula (89):

$$(\mu\mu|\mu\mu) = 3.29_4 \, Z_\mu \qquad [2]$$

In order to determine the dependence of the valence-state ionization potentials on q_μ they plotted I_μ against Z_μ for the isoelectronic series C, N^+, O^{2+}, F^{3+}. They found the curve accurately parabolic and used it to evaluate I_μ.

In 1963 Brown and Harcourt (90) extended the VESCF method to systems containing both σ and π electrons. Their targets were A_2Y_4 type molecules and, in particular, the explanation of the bond lengths of central bonds involving delocalization of the Y lone-pair electrons (of $2\underline{p}$ or $3\underline{p}$ type with axes in the molecular plane) into an antibonding σ orbital between the A atoms. Several characteristics of later methods appear in Brown and Harcourt's treatment. One of these is the determination of Hückel parameters from fitting the results of Roothaan SCF calculations, actually VESCF, in their case. Another is their prediction of significant lone-pair delocalization.

In their Pariser-Parr type calculations, the σ-electron repulsion integrals are computed through the π-electron Mataga and Nishimoto formula (91) using the relationship

$$(\gamma_{\mu\nu}{}^\sigma) = \frac{(\gamma_{\mu\nu}{}^\sigma)_T}{(\gamma_{\mu\nu}{}^\pi)_T} \, (\gamma_{\mu\nu}{}^\pi) \qquad [3]$$

where the index T means Slater theoretical values. For $\mu = \nu$ they used the same ratio but computed $(\gamma_\mu{}^\pi)$ from Paolini's formula (89). We refer to Brown and Harcourt (90) for the details of obtaining the valence-state ionization potentials necessary for the computation of the diagonal core terms (α). The β were computed from the formulas

$$\beta^{\pi}_{\mu\nu} = -\tfrac{1}{2}S^{\pi}_{\mu\nu}\,(I^{\pi}_{\mu} + I^{\pi}_{\nu}) \text{ and } \beta^{\sigma}_{\mu\nu} = -\tfrac{1}{2}S^{\sigma}_{\mu\nu}\,(I^{\sigma}_{\mu} + I^{\sigma}_{\nu}) \quad [4]$$

Neutral-atom penetration integrals were omitted in computing the core coulomb integrals.

This interesting method should be useful in treating more general systems. Its handling of orbital exponents provides it with a kind of versatility which is perhaps not attainable otherwise (see p. 94).

In a later paper Brown and Harcourt (92) reported more detailed calculations on N_2O_4, including all σ and π electrons. They made alternate π and σ VESCF iterations until both sets of molecular orbitals reached self-consistency. Only the delocalization from $2p\bar{\pi}$ orbitals into the antibonding NNσ orbitals turns out to be important. ($2p\bar{\pi}$ orbitals have their axis perpendicular to the double-bonding π orbital.)

In recent papers Brown and Peel (93) extended the VESCF MO method to molecules containing atoms from the second row of the periodic system.

Allinger et al. (94) carried out an extensive VESCF CI treatment of the electronic spectra of unsaturated hydrocarbons allowing for the inductive effect of alkyl groups.

A somewhat simpler procedure was used by Allinger and Tai (95), who proposed one of the first methods dealing with σ-electron systems in a paper which was concerned with the "size of the lone pair on nitrogen." They computed the energies of a system composed of an ammonia molecule and a helium atom approaching either along the threefold axis or in the direction of a N–H bond. They found that the first approach is much more favorable than the second one and that the charge distribution corresponding to a lone pair is smaller than, say, a methyl group.

In their preliminary Hückel calculations they obtained the α_X for an atom X from a formula proposed by McWeeny and Peacock (96):

$$\delta\alpha_X = \left\{(A_X - A_C) + \tfrac{1}{2}[Z_X(I_X - A_X) - (I_C - A_C)]\right\}/(-4.79) \quad [5]$$

where the subscript C refers to carbon, I and A are the respective valence-state ionization potentials and electron affinities, and Z_X is the number of electrons contributed by the orbital centered on X. $\delta\alpha_X$ is expressed in units of $\beta_0 = -2.39\underline{ev}$, the Hückel β for benzene. The resonance integrals were computed from one of Mulliken's formulas (97):

$$\beta_{\underline{ij}} = \frac{\frac{1}{2}(I_{\underline{i}}+I_{\underline{j}})\ A_{\underline{ij}}S_{\underline{ij}}/(1 + S_{\underline{ij}})}{I_{\underline{o}}A_{\underline{o}}S_{\underline{o}}\ /\ (1 + S_{\underline{o}})}\ \beta_{\underline{o}} \qquad [6]$$

where $\underline{i}$ and $\underline{j}$ designate neighboring atoms and $\underline{o}$ a π orbital of benzene. $A_{\underline{ij}}$ is 0.773 for σ-type and 1 for π-type overlap.

In their Pariser-Parr calculations the following choice of integrals was made: $N\underline{sp}^3$ hybrids and $H1\underline{s}$ orbitals were used as basis functions and the one-center one-orbital coulomb integrals were estimated in Pariser's manner. Differential overlap was neglected. Two-center coulomb integrals were computed from an extrapolation formula in Pariser and Parr's manner.

The core integral $\alpha_{\underline{p}}$ was computed with the Goeppert-Mayer and Sklar Hamiltonian

$$H^{core}(\underline{i}) = T(\underline{i}) + U_p(\underline{i}) + \sum_{q \neq p} U_q(\underline{i}) + \sum_r\ U_r^*(\underline{i}) \qquad [7]$$

where U_p, U_q and U_r^* are, respectively, the potential energy operators in the field of the charged atoms p, q, and the uncharged atoms r which constitute the core. Then

$$\gamma_p = (\chi_p|H^{core}|\chi_p) = -I_p - \sum_{q \neq p} n_q(q:pp)\ -\ \sum_r (r:pp) \qquad [8]$$

where n_q is the charge on atom q, $(q:pp)$ is a penetration integral for a charged atom q, and $(r:pp)$ for a neutral atom r. The former were replaced by the negatives of the corresponding electron repulsion integrals and were scaled down the same way. The Hückel β values were multiplied by -2.39 to convert their values into electron volts. Some features of this method reappear in methods later proposed by other authors.

2. The CNDO method*

The CNDO method due to Pople, Santry, and Segal (98) is, at present, the most popular among the PPP-type methods applied

---

*Note added in proof: Approximate Molecular Orbital Theory, J. A. Pople and D. L. Beveridge (McGraw Hill, New York, 1970), contains a thorough treatment of CNDO methods and applications.

to σ-electron systems and is amenable to a variety of applications and improvements. Pople et al. originally proposed two different approximations: "complete neglect of differential overlap" (CNDO) and "neglect of diatomic differential overlap" (NDDO).

In the CNDO method, first, the usual PPP approximations are made; that is, both overlap integrals and differential overlap are neglected. It is most important to note that some of these will concern differential overlap between atomic orbitals on the same atom.

At this stage Pople et al. observed that the "theory is not invariant under a rotation of local axes or under hybridization." It is easy to see that, for example,

$$2(2px'_A, 2py'_A \,|\, 2s_B, 2s_B) = (2py_A, 2py_A \,|\, 2s_B, 2s_B)$$
$$-(2px_A, 2px_A \,|\, 2s_B, 2s_B) \qquad [9]$$

where x and y axes are rotated by 45° with respect to the x' and y' axes. The left-hand side would vanish under zero-differential overlap approximation but the right-hand side would not.

Furthermore, since sets of hybrid and the related "pure" atomic wavefunctions are connected by an orthogonal transformation, the results should be unchanged by such a transformation. However, the neglect of differential overlap in integrals taken over pure orbitals would not, in general, make vanish an integral taken over hybrid orbitals based on the former.

Invariance is restored in both cases if the following new approximation is adopted: "The electron interaction integrals $\gamma_{\mu\nu}$, are assumed to depend only on the atoms to which the atomic orbitals χ_μ and χ_ν belong and not on the actual type of orbital." This is the most characteristic approximation of the CNDO method and is tantamount to taking the average repulsion between atoms in valence orbitals on atoms A and B. Actually, Pople and Segal (99) found that this average is very close to the electron repulsion integral for $2s$ electrons, and used this value in their calculations:

$$\gamma_{AB} = \int s_A^2(1) \frac{e^2}{r_{1,2}} s_B^2(2) \, d\tau_1 \, d\tau_2 \qquad [10]$$

The matrix components in the self-consistent version of the PPP method are (100), if we make only the S = O and ZDO approximations (as in the π-electron case):

$$F_{\mu\mu} = H_{\mu\mu}^{core} + \tfrac{1}{2}P_{\mu\mu}\gamma_{\mu\mu} + \sum_{\sigma \neq \mu} P_{\sigma\sigma}\gamma_{\mu\sigma} \qquad [11]$$

$$F_{\mu\nu} = H_{\mu\nu}^{core} - \tfrac{1}{2}P_{\mu\nu}\gamma_{\mu\nu} \; (\mu \neq \nu) \qquad [12]$$

Here μ, ν, σ stand for atomic orbitals, $P_{\mu\mu} \equiv q_\mu$ is the electronic charge density in μ, $P_{\mu\nu}$ is the bond order of bond $\mu-\nu$, and the γ are one-orbital and two-orbital coulomb integrals. Then, if we apply the third approximation of Pople, Santry, and Segal (average coulomb integrals) these expressions become:

$$F_{\mu\mu} = H_{\mu\mu}^{core} - \tfrac{1}{2}P_{\mu\mu}\gamma_{AA} + P_{AA}\gamma_{AA} + \sum_{B \neq A} P_{BB}\gamma_{AB} \qquad [13]$$

$$F_{\mu\nu} = H_{\mu\nu}^{core} - \tfrac{1}{2}P_{\mu\nu}\gamma_{AB} \; (\mu \neq \nu) \qquad [14]$$

In the last equations

$$P_{BB} = \sum_\nu {}^B P_{\nu\nu}$$

is the total valence-electron density on atom B, and γ_{AB} becomes γ_{AA} in the last equation if both μ and ν are on atom A. (P_{AA} contains $P_{\mu\mu}$, and $P_{AA}\gamma_{AA}$ represents all the repulsions on an electron in μ exerted by other electrons on A. These latter were contained in $\sum_{\sigma \neq \mu} P_{\sigma\sigma}\gamma_{\mu\sigma}$ but are not contained in $\sum_{B \neq A} P_{BB}\gamma_{AB}$.)

Two further approximations become necessary if we write out the core integrals explicitly. For the diagonal core integral

$$H_{\mu\mu}^{core} = (\mu|-\tfrac{1}{2}\nabla^2 - V_A|\mu) - \sum_{B \neq A} (\mu|V_B|\mu) =$$

$$U_{\mu\mu} - \sum_{B \neq A} (\mu|V_B|\mu) \qquad [15]$$

where, as usual $U_{\mu\mu}$ represents the interaction of a valence electron with its "own" core, and the sum its interaction with all other cores. For the off-diagonal core integrals we have

$$H_{\mu\nu}^{core} = U_{\mu\nu} - \sum_{B \neq A} (\mu|V_B|\nu) \qquad [16]$$

When μ and ν are on the same atom, $U_{\mu\nu} = 0$ if pure atomic orbitals are used, and Pople et al. neglect $(\mu | V_B | \nu)$ which contains differential overlap.

Then $\quad H_{\mu\mu}^{core} = U_{\mu\mu} - \sum_{B \neq A} V_{AB} \; (\mu \text{ on } A)$ [17]

with $V_{AB} \equiv (\mu | V_B | \mu)$,

$\quad H_{\mu\nu}^{core} = 0 \quad (\mu \neq \nu, \text{ both on } A)$ [18]

Finally if μ and ν are on two different atoms, the core term will be taken proportional to the overlap integral,

$$H_{\mu\nu}^{core} = \beta_{\mu\nu} = \beta_{AB}^{O} \, S_{\mu\nu}$$ [19]

where β_{AB}^{O} will depend only on the nature of atoms A and B. The latter is determined empirically or chosen so as to give agreement with the results of more elaborate calculations (mainly LCAO MO SCF). These are the basic approximations of the CNDO method.

To determine the actual values of the integrals, Pople et al. computed the coulomb integrals from Roothaan's formulas using Slater s orbitals. The results turned out to be very close to their average values over an atom A or for a pair of atoms A and B. They did not use Pariser and Parr's IA (ionization potential-electron affinity) formula, or any of the conventional methods of estimating the two-center coulomb integrals. The "own-atom" part of the diagonal core integral is computed from observed atomic energy levels. For an atom X with m electrons in 2s and n electrons in 2p orbitals, for example,

$$E(X, 2s^m \, 2p^n) = m \, U_{2s,2s} + n \, U_{2p,2p}$$
$$+ \tfrac{1}{2}(m+n)\,(m+n-1)\,\gamma_{XX}$$ [20]

If we write a similar expression for the respective positive ion, the difference can be equated to the experimental ionization potential and the U can be calculated with only γ_{XX} needed in addition to the latter. The atomic energy levels are taken to be a weighted average of all states arising from the given configuration (not in every case equal to a valence state energy). Mean values are used for the ionization potentials too. A tabulation of these is given by Pople and Segal (99). The intermolecular part of the diagonal core term (V_{AB}) is also computed with s orbitals and the point-charge approximation is used so that

$$V_{AB} = \int s_A{}^2(1) \frac{Z_B}{r_{1B}} d\tau,$$

where Z_B is the core charge.

As we said above, the off-diagonal core terms are taken proportional to the overlap integrals. The proportionality factor $\beta_{AB}^O = \frac{1}{2}(\beta_A^O + \beta_B^O)$ where the β_A^O are determined empirically to give agreement with full LCAO MO SCF calculations. ($-\beta_A^O$ is 9, 9, 13, 17, 21, 25, 31 and 39 ev for H, Li, Be, B, C, N, O and F respectively.)

In the same publication, Pople, Santry, and Segal (98) also proposed a modified and very probably much better but more complicated method, called the NDDO (neglect of diatomic differential overlap) approximation. In this modified scheme differential overlap in two-electron integrals is neglected only for orbitals on different atoms, that is, $(\mu\nu|\gamma\sigma) = 0$ except if μ, ν are on A and γ, σ on B. To be consistent then, the terms $\sum\limits_{B\neq A} (\mu|V_B|\nu)$ elements in the off-diagonal core are also retained.

The NDDO approximation has been thoroughly examined and justified recently by Cook et al. (101).

Pople and Segal (99) tested the CNDO method on a large number of small molecules, computing electronic charge distributions, energies, equilibrium configurations, and vibrational force constants. They found that if the β_A^O were calibrated to give agreement with full SCF calculations on diatomic hybrids using a similar basis set, the wavefunctions and electron populations were about the same as in the full LCAO SCF method. The dependence of the energy on internuclear distance was not satisfactory. (This is expected to be very sensitive to the neglect of overlap and differential overlap.) The bond angle dependence of the energies gave much better results.

We reproduce here some of the charges (q) obtained by Pople and Segal with unhybridized orbitals.

Staggered ethane   z axis in the C–C direction. 14 valence electrons.

C	2s	1.0419
	$2p_x$	1.0071
	$2p_y$	1.0071
	$2p_z$	1.0438
H	1s	0.9666

Water

x axis according to the twofold symmetry axis, z perpendicular to the molecular plane. 8 valence electrons.

O	2s	1.6920
	$2p_x$	1.3666
	$2p_y$	1.0577
	$2p_z$	2.000

H	1s	0.9418

Ammonia

z axis according to the threefold symmetry axis. 8 valence electrons.

N	2s	1.4414
	$2p_x$	1.0387
	$2p_y$	1.0387
	$2p_z$	1.6224

H	1s	0.9530

Very similar methods were proposed by Kaufman (102). She considered successive approximations which were the CNDO, and the NDDO with or without the inclusion of two-center exchange integrals and the inclusion of the two-center hybrid integrals. She also proposed the use of valence-state ionization potentials for obtaining one-center core integrals modified according to the electronic charge (q) of the given orbital. Applica= ?

3. Modified CNDO methods

The CNDO method predicted molecular geometries, bending force constants, and rotational barriers in fair agreement with experimental values for small polyatomic molecules. Its principal failure was that for diatomic molecules it led to calculated bond lengths which were too short and binding energies which were too large. According to its authors this is "because of a 'penetration' effect in which electrons in an orbital on one atom penetrate the shell of another leading to net attraction."

Thus, in a later paper, Pople and Segal (103) modified the CNDO method in two ways. The diagonal matrix element [13] can be rewritten as

$$F_{\mu\mu} = U_{\mu\mu} + (P_{AA} - \tfrac{1}{2}P_{\mu\mu})\, \gamma_{AA}$$

$$+ \sum_{B \neq A} (P_{BB} - Z_B)\, \gamma_{AB} + \sum_{B \neq A} (Z_B \gamma_{AB} - V_{AB}) \tag{21}$$

where Z_B is the core charge of atom B. Then the last term can be considered as the contribution of penetration integrals to $F_{\mu\mu}$. These lead to bonding energies even when the relating bond orders are zero. Thus Pople and Segal (103) put

$$V_{AB} = Z_B \gamma_{AB}$$

which amounts to neglecting the last term in [21]. This is justified by the cancellation of errors due to this approximation and the neglect of overlap distributions, and leads to better agreement with experiment.

The second modification concerns the local core matrix elements $U_{\mu\mu}$. These were computed from

$$-I_\mu = U_{\mu\mu} + (Z_A - 1)\, \gamma_{AA}$$

Since it would have been equally possible to use the electron affinities instead of the ionization potentials:

$$-A_\mu = U_{\mu\mu} + Z_A \gamma_{AA}$$

the authors decided to use the average of the $U_{\mu\mu}$ obtained from these two expressions. Then

$$-\tfrac{1}{2}(I_\mu + A_\mu) = U_{\mu\mu} + (Z_A - \tfrac{1}{2})\, \gamma_{AA} \tag{22}$$

and it is seen that in this way the Mulliken electronegativities enter the $F_{\mu\mu}$ which become:

$$F_{\mu\mu} = -\tfrac{1}{2}(I_\mu + A_\mu) + [(P_{AA} - Z_A) - \tfrac{1}{2}(P_{\mu\mu} - 1)]\, \gamma_{AA}$$

$$+ \sum_{B \neq A} (P_{BB} - Z_B)\, \gamma_{AB} \tag{23}$$

while

$$F_{\mu\nu} = \beta^{O}_{AB}\, S_{\mu\nu} - \tfrac{1}{2}P_{\mu\nu}\, \gamma_{AB} \tag{24}$$

If the orbital contains only one electron, $P_{\mu\mu} = 1$; and if the net charges are zero on all atoms, $P_{AA} = Z_A$, $P_{BB} = Z_B$, then only the electronegativities remain. This is a slightly different way of making the latter appear in the diagonal matrix element than the one found by Klopman (see p. 38).

Pople and Segal called this version of the CNDO method the CNDO/2 method. It is now widely used. They also extended it to open-shell configurations by using different Slater determinants for molecular orbitals with α and β spins. This method, among others, makes it possible to define a spin-density matrix

$$Q_{\mu\nu} = P_{\mu\nu}{}^{\alpha} - P_{\mu\nu}{}^{\beta}$$

In the same paper, Pople and Segal applied the CNDO/2 method to a series of linear and nonlinear AB_2 type molecules. They used fixed bond lengths and computed the total energy and wavefunctions as functions of the BAB angle. This amounts to a derivation of Walsh's rules (23) and makes pleasant reading for spectroscopists. (See their Table II (103) and the discussion of Hoffman's method (22, 24—27) in Chapter 1.)

One-center exchange integrals are neglected in the CNDO/2 method just as in CNDO/1, and this leads to poor predictions of singlet-triplet separations. They should be better in the NDDO approximation.

In a subsequent paper, Pople and Gordon (104) used the CNDO/2 method to calculate charge distributions and dipole moments of a series of simple organic molecules. The originality of this work is the use of "standard geometrical models" rather than molecular geometries fixed on experimental data in order to see how the geometry is influenced by the electronic structure. "For example, in the theory of alternation of bond lengths in polyenes, it is better to start with a theory of the electronic struc-ture based on equal bond lengths and then discover the "driving force" causing alternation (alternating bond order), rather than starting with a geometrical model with the alternation already built in."

They directed their efforts mainly toward substituent effects. Both mesomeric and inductive effects turn out to be alternating with increasing distance from the substituent. (See their discussion of fluorine-substituted compounds.)

Santry and Segal (105) extended the CNDO/2 method to molecules containing elements of the second row of the periodic table (sodium to chlorine) in order to examine the role of 3d orbitals. They tried three different basis sets. (1) "sp": 3d functions are excluded from the basis set (their orbital exponents are taken for zero). (2) "spd": the orbital exponents of the 3d functions are taken to be the same as those of the 3s and 3p orbitals. This introduces a significant contraction of the d orbitals. (3) "spd'": an intermediate value is used for the 3d orbital exponents.

The degree of contraction that the 3d electrons undergo in molecules is closely related to their bonding ability and there is considerable uncertainty about it. Santry and Segal found that the intermediate basis set (with moderate contraction) leads to the best agreement with experiment as to molecular geometries, dipole moments, and bending force constants. The dipole moments are sensitive to 3d contributions but molecular geometries are less so. The authors list separately the contributions to the dipole moments from molecular polarity, from sp lone-pair and from pd lone-pair polarities. The d electrons have a significant effect on charge distribution and bonding conditions. Tables 4 and 5 relating to SF_6 illustrate this (105).

Brown and Burden (106) published a paper on the "optimum parametrization" of the CNDO method deriving the $U_{\mu\mu}^{AA}$ for 2p and H1s orbitals empirically by a least-squares fitting of calculated to observed dipole moments. The $U_{2s,2s}^{AA}$ and resonance integrals were found to have little effect on the calculated moments and the optimum values of the $U_{2p,2p}^{AA}$ and $U_{1s,1s}^{AA}$ were close to Pople, Santry, and Segal's CNDO/2 values except in cases of large bond polarity or atomic dipole terms.

Bloor, Gilson, and Billingsley (107) compared the dipole moments calculated by the CNDO/2, EH, and iterative EH methods for HF, H_2O, NH_3, CO, H_2CO, $HCONH_2$, HCOOH, HCOF, and sydnone. They carefully considered terms due to the Mulliken overlap population, atomic dipole (asymmetry of lone pairs), and overlap moment (homopolar dipole contribution). They found that the EH method yields moments that are too high. The IEH and CNDO/2 values are usually good, the latter leading to better values for high dipole moments.

There is an interesting possibility of introducing a further approximation, different from CNDO and CNDO/2 as well as from NDDO. This is the "intermediate neglect of differential overlap" (INDO) which was proposed by Pople, Beveridge, and Dobosh (108) and by Dixon (109). It consists of applying the zero differential overlap approximation in all electron interaction integrals except those involving one center only; that is, $(\mu\nu|\mu\mu) \neq 0$ but $(\mu\nu|\lambda\lambda) = 0$ if μ and ν are on the same center but λ is on a different center.

Table 4. Contracted s̲p̲d̲ bond-order matrix for SF_6

F	S				
	3s̲	3p̲z	3p̲x	$3dz^2$	3d̲xz
2s̲	0.138	0.282	0	0.274	0
2p̲z	0.382	0.576	0	0.424	0
2p̲π	0	0	0.133	0	0.343

The s̲p̲ bond-order matrix is very similar except for the 3d̲ elements. The fluorine lies along the z axis. After Santry and Segal (105).

Table 5. Charge distribution between the atomic orbitals of sulfur and fluorine in SF_6 in the s̲p̲ and s̲p̲d̲ approximations

Orbital	F electron distribution		S electron distribution	
	s̲p̲d̲	s̲p̲	s̲p̲d̲	s̲p̲
s̲	1.868	1.927	1.104	1.325
p̲x	1.919	1.986	0.672	0.871
p̲z	1.467	1.444	0.672	0.871
d̲e$_{1g}$			0.514	
d̲t$_{2g}$			0.272	
Total	7.173	7.343	4.964	3.938

The fluorine atoms are situated on the molecular axes. After Santry and Segal (105).

The latter would not be zero in the NDDO approximation. This saves much of the extra work involved with the NDDO approximation and conserves most of its advantages. It improves singlet-triplet separations and spin densities. Gordon and Pople (110) computed equilibrium geometries for many molecules containing C, N, O, and F using the INDO scheme and neglecting penetration integrals.

Newton, Ostlund, and Pople (111) recently worked out a rapid and general method for the approximate calculation of two-, three-, and four-center potential energy integrals with rotational invariance maintained. This is the "projection of diatomic differential overlap" (PDDO) method which consists of a least-squares projection of all two-center, one-electron distributions onto one-center distribution basis functions.

The CNDO method and its modified forms are widely used for treating specific problems of molecular properties.

4. Extensive use of atomic energy data in obtaining parameters

One of the difficulties we have to face when we turn from π-electron problems to all-valence electron problems is the need to consider explicitly the presence of more than one atomic orbital on the same atom. This introduces a number of new integrals involving two or more AO on the same atom. Their values, which should enter a semiempirical calculation, are difficult to estimate. The original CNDO method takes the averages of the coulomb integrals and neglects the integrals containing differential overlap. The NDDO method takes averages for both. Klopman (112) uses atomic spectral data to cover all atomic integrals on the same atom.

Let Φ_Λ be the wave function relating to a configuration Λ of a given atom, expressed in form of a Slater determinant. The total Hamiltonian is, for the n valence electrons $(1, 2, \ldots, i, j, \ldots, n)$:

$$H = \sum_i H_{core}(i) + \sum_{ij} \frac{e^2}{r_{ij}} \qquad [25]$$

where the core only contains the nucleus and the electrons of the inner shells. The energy is then

$$E_\Lambda = \int \Phi_\Lambda^* H \Phi_\Lambda dt$$

or $$E_\Lambda = \sum_q I_q + \tfrac{1}{2} \sum_{qr} (J_{qr} - K'_{qr}) \qquad [26]$$

where, as usual, I_q is the core integral for an atomic spin orbital χ_q, J_{qr} is the coulomb integral between pairs of spin orbitals and K'_{qr} is the exchange integral between spin orbitals having the same spin. Since the atomic states are intended for use in molecular calculations, we have to consider those valence states which intervene in such calculations. Klopman classifies the atomic states according to the values of M_s^2, the component of the total spin angular momentum in a given direction. If there are several states issued from the same atomic configuration having the same M_S^2 value, their center of gravity, or "barycenter" is taken. For an $\underline{s}^2 p^2$ configuration with $M_S = 0$ for example, we take

$$(5{}^1D + {}^1S + 3{}^3P)/9$$

and for the same configuration with $\pm M_S = 1$, simply ${}^3P$.

For atom x, I_q is equated to a constant B_x^l where l is the azimuthal quantum number. The total interaction energy between two electrons with the same spin and the same azimuthal quantum numbers l and l', is with Klopman's notations:

$$J_{ll'} - K_{ll'} = A^+_{ll'}$$

and between two electrons of opposite spin:

$$J_{ll'} = A^-_{ll'}$$

Klopman then introduces the simplification that $A^+_{ll'}$ and $A^-_{ll'}$ remain constant provided the two electrons belong to the same shell even if l and l' are not equal. Then he defines A^+_x as the mean value of the $A^+_{ll'}$ and A^-_x as the mean value of the $A^-_{ll'}$ for all the values of l. The energy of the atomic states then becomes

$$E = \sum_i B_x^l + \tfrac{1}{2} \sum_{ij} A_x^+ \delta_{ij} + \tfrac{1}{2} \sum_{ij} A_x^- (1 - \delta_{ij}) \qquad [27]$$

where δ_{ij} is equal to 1 if the two spins are parallel and to 0 if they are antiparallel.

In the example $(\underline{s}^2 p^2$ with $M_S = 0)$ we then have:

$$E = 2B^s + 2B^p + 4A^- + 2A^+$$

since the sums are unrestricted in [27]. For the configuration $\underline{s}^2 p^2$ with $\pm M_S = 1$

$$E = 2B^s + 2B^p + 3A^- + 3A^+$$

Values of B and A can be computed from experimental transition energies which Klopman takes from the Tables of Atomic Energy Levels of C. E. Moore (113). He also used electron affinity values from Pritchard (114) and Edlén (115). For carbon, for example,

$$B^s = -49.884 \text{ ev} \qquad B^p = -42.696 \text{ ev}$$
$$A^- = 11.144 \text{ ev} \qquad A^+ = 10.144 \text{ ev}$$

There is an extensive tabulation in (112).

Klopman makes the interesting remark that if all electrons have the same azimuthal quantum number, then,

$$B_X^{\ 1} = B_X \text{ (constant)}$$

and if the exchange integrals are neglected,

$$A^+ = A^- = A_X = \text{constant}$$

Then the energy expression under [27] simplifies to

$$E = \sum_i B_X + \tfrac{1}{2} \sum_{ij} A_X \qquad [28]$$

If q is the number of electrons

$$E = qB_X + \frac{q(q-1)}{2} A_X \qquad [29]$$

and if we introduce

$$a = B_X - A_X/2 \text{ and } b = A_X/2$$

we obtain

$$E = aq + bq^2 \qquad [30]$$

This is the same as the formula of Iczkowski and Margrave (116) which is the starting point for the orbital electronegativity concept, first introduced by Mulliken (18) and developed in recent times by Hinze and Jaffe (21), and Hinze, Whitehead, and Jaffé (117).

For just one orbital the last equation is as exact as equation [27] since the $B_X^{\ 1}$ are of course the same for the two electrons, and $K = 0$ since they must have anti-parallel spins.

This leads to a definition of electronegativity in energy terms. It now appears in the diagonal terms of the matrices. It will change from molecule to molecule according to the coefficient that the given atomic orbital has in the molecular orbitals.

In his second paper (118), Klopman extended his method to diatomic molecules. Like Pople et al. he neglects differential overlap and disregards nonbonded interactions as well. The characteristic difference between the two methods is the extent to which they utilize atomic spectral data. Pople et al. use them only to obtain the one-center part of the diagonal core term ($U_{\mu\mu}$) while Klopman takes the one-center electron interaction integrals from atomic spectral terms as well. The one-center coulomb integrals are averaged for the atoms in both methods.

Since all the atomic integrals are derived from spectral data in Klopman's method, he only has to determine the two-center coulomb integrals and the two-center core integrals (the β).

For the former he took the point-charge approximation with $\underline{e}^2/r$ as a trial function. This was adequate except if the bond was formed by $\underline{s}$ electrons. In the latter case, a modified point-charge formula gave better results:

$$e^2/\sqrt{r^2 + (\rho_x + \rho_y)^2}$$

where the ρ represent the radii of the appropriate $\underline{s}$ orbitals. The latter are tabulated for several elements in (118). In this way, the charges are no longer located at nuclear centers, and mutual polarization is taken into account to some extent. As for the β, Pauling's geometrical mean formula was used,

$$\beta_{AB} = (\beta_{AA} \, \beta_{BB})^{1/2}$$

In Klopman's third paper (119) this method is generalized to polyatomic molecules. LCAO molecular orbitals are introduced. Integrals containing differential overlap are neglected only between AO on different atoms and those which are on the same atom are taken from atomic spectra as indicated above. Klopman pointed out that equating these to zero amounts to equating the interaction between two electrons with the same spin and with opposite spins (cf. equation [39]). This causes a problem if there are more than two bonding electrons. The difference is significant, as seen from atomic spectral data.

An interesting feature of this method is the minimization of the energy with respect to the charges (q) of the orbitals. The potential around each atom in each orbital becomes the same:

$$\frac{\partial E_{total}}{\partial q_x{}^i} = \frac{\partial E_{total}}{\partial q_y{}^i}$$

where $q_x{}^i$ and $q_y{}^i$ are the charge densities due to electron i in atomic orbital x and y. This is in keeping with the concept of molecular electronegativity equalization due to Sanderson (120) which Ferreira (121) had previously introduced into diatomic calculations. Klopman calls the molecular orbitals obtained in this way equipotential orbitals.

More recently Dewar and Klopman (122) presented a modified scheme whose aim is to compute heats of formation with
"chemical" accuracy (about ±1 kcal/mole). (See also the chapter "Energetics with Sigma Molecular Orbital Theory" in SMOT.)
This method is characterized by the same extensive use of atomic spectral data as that which was described above. The local
axes on the atoms are chosen in such a way as to make all two-
center three-orbital integrals involving overlap between pairs of
p atomic orbitals vanish through symmetry, as well as most
four-orbital integrals. The remainder of the latter are neglected
but it can be shown that this does not seriously affect the invariance of the calculations to the choice of coordinate axes. In this
way, different integrals can be used for s and p atomic orbitals
of a given center and there is no need to average them. One-center exchange integrals are also included. Dewar and Klopman call
this the "Partial neglect of diatomic differential overlap" (PNDDO)
approximation.

They found it equally necessary to use a more finely adjust-
able formula for the two-center core integrals β_{kl}:

$$\beta_{kl} = (\beta_{kl})_0 \, S_{kl}(I_k + I_l) \, [r_{kl}^2 + (\rho_k + \rho_l)]^{-1/2} \qquad [31]$$

where I_k and I_l are valence-state ionization potentials of the
atomic orbitals involved, r_{kl} is the internuclear distance, $(\beta_{kl})_0$
is an empirical parameter which is the same for all valence orbitals of a given atom, S_{kl} is the overlap integral, and ρ_k and ρ_l
are obtained from the formula used to adjust the two-center coulomb integrals. The β therefore depend on the "magnitude of the
overlap cloud" through the S_{kl}; on the mean of the binding energies of the respective atomic orbitals; and on the distance between the overlap cloud and the nuclei of the atoms.

In the simplest case their two-center coulomb integrals are:

$$\gamma_{kl} = \underline{e}^2 \, [r_{kl}^2 + (\rho_k + \rho_l)^2]^{-1/2}$$

where the ρ are chosen to make γ_{kl} approach the corresponding
one-center integral for $r_{kl} \to 0$.

As another innovation they treat internuclear repulsion as a
parameter to allow for the effects of orbital contraction.

They use thermochemical data of the simplest molecules to
fix the parameters rather than using the results of Hartree-Fock
calculations as did Pople (98). We refer the reader to the original paper of Dewar and Klopman (122) for the details of this
somewhat elaborate method.

Baird and Dewar applied the PNDDO scheme to the ground states of σ-bonded molecules. They studied strained small-ring molecules (123) and conjugated molecules (124). For the latter they compared the results of the all-valence electron calculations to those of π-electron calculations. They found a good correlation between the π energy levels obtained in the two methods as well as the bond orders. They found that hyperconjugation in molecules such as propene and toluene is negligible but that it is important in carbonium ions. (For another recent discussion of hyperconjugation see also Flurry (125).) Dewar et al. (126) studied hydrocarbon radicals and radical ions by a modified closed-shell method.

Oleari et al. (127) used valence-state energies in their semiempirical valence electron method which is a different way of using atomic spectral data than Klopman's. Instead of Pariser and Parr's IA formula they used, for the all-valence electron case for an atom with $\underline{s}$ and $\underline{p}$ electrons:

$$E = C + \sum_i n_i U_{ii} + \tfrac{1}{2} \sum_i \sum_{j \neq i} n_i n_j \gamma_{ij} + \tfrac{1}{2} \sum_i n_i(n_i - 1)\gamma_{ii}$$

[32]

where the n_i are orbital occupation numbers and the sums contain the U_{ss}, U_{pp} and γ_{ss}, γ_{sp}, and γ_{pp} respectively for all the orbitals of the atom. All the parameters occurring in [32] are determined from atomic spectra. They also took into account variation with atomic charge and discussed the importance of the constant C. The latter represents the energy of the "core state" (all valence electrons removed) and should not be set equal to zero since it would then actually enter into the values of the parameters which we are to determine.

Sichel and Whitehead (128) introduced these parameters into the CNDO method, evaluating the one-center core and electron repulsion matrix elements (the "atomic" parameters) from atomic valence-state energies instead of Slater 2$\underline{s}$ orbitals. They used atomic parameters based on the valence state energies given by Hinze and Jaffé (21, 129). They averaged the γ_{ij} to obtain the γ_{AA} for an atom A and adjusted the core elements U_{ss} and U_{pp} after the averaging of the γ_{AA}.

In a second paper Sichel and Whitehead (130) used the above parametrization together with the Mataga (131) or Ohno (132) formula for diatomic electron-repulsion integrals and calibrated the bonding parameters (the β^0) to match the bonding energies of diatomic hydrides. β_H^o was chosen to give the correct dissociation energy for the hydrogen molecule. They found better agreement with experiment than either the CNDO/2 or the EH methods for bonding energies.

Baird, Sichel, and Whitehead (133) made an attempt to introduce the electronegativity equalization concept into molecular orbital theories through localized bonds in molecules.

Harris (134) in a recent note made critical comments on the dependence of Hückel parameters upon the charge distribution. He has shown that the minimum-energy molecular orbitals are not the same as the results of iterative solutions of the usual Hückel equations. There will be, no doubt, further developments (see, for example, Cusachs and Politzer (59, 60); and see Chapter 4).

We should like to mention at this point a very interesting contribution by Saturno (135) who, by replacing $r_{\underline{ij}}^{-1}$ by $(r_{\underline{i}} + r_{\underline{j}})^{-1}$, derived the relation

$$2\zeta/(4l+5) = I - A \qquad [33]$$

where ζ is the Slater orbital exponent and l is the effective principal quantum number. For $2\underline{p}$ orbitals this is equal to

$$Z_p/9 = I_p - A_p$$

where Z_p is the Slater effective nuclear charge. Coulson (136) gave the rigorous derivation of Saturno's formula. This is rather useful for the calculation of one-center one-orbital coulomb integrals but it can be used for obtaining two-center coulomb integrals also (136).

Pearson and Gray (137) examined the partial ionic character of metal-chlorine bonds and criticized the concept of electronegativity equalization.

Ferreira (138) made an important study on bond energies, hybridization states, and bond ionicities for diatomic molecules.

Peters (139) examined the role of inner shells and found that their effect consists in increasing the ionization energy of the valence atomic orbitals of the atom in the molecule but that they have no significant effect on bond energies.

The early work of Pearson (140) on the calculation of ionic resonance energies and Arnold's (141) smoothed potential theory of chemical binding should be acknowledged.

Veillard and Berthier (142) used a mixed method to study pyridine-like molecules. The diagonal matrix elements were obtained from atomic spectral data through the SCF equations, and the nondiagonal ones through the Wolfsberg-Helmholtz formula (19).

Labarre et al. (143) studied the $(\sigma + \pi)$ electronic structures of substituted borazines and boroxines.

Tinland (144) made a CNDO study of steric effects in biphenyl.

Kroner and Bock (145) examined d-orbital effects in π-electron systems containing silicon.

Many others should be mentioned.

5. Methods using Mulliken's approximation

In one of the earliest σ-electron methods, Pohl, Rein, and Appel (146) concerned themselves with hydrogen halides, taking into account explicitly only the two bonding electrons, putting all other electrons into the core. This way they avoided the complications arising from the presence of more than one valence orbital on the same atom. Their method is interesting, however, in several respects.

They used the complete Roothaan equations with a single configuration without outright neglecting differential overlap. Instead they applied Mulliken's approximation. For electron-electron and electron-core interactions they used the simple point-charge approximation.

Then, the diagonal core integrals are of the form (if 1 and 2 are the two atomic orbitals involved):

$$H_{11} = \langle 1 | - \tfrac{1}{2} \nabla_1^2 + V_1 + V_2 | 1 \rangle = -I_1 + \langle 1 | V_2 | 1 \rangle =$$
$$-I_1 - R_{12}^{-1} \qquad [34]$$

where I_1 is the appropriate valence-state ionization potential and R_{12} is the internuclear distance. The nondiagonal core terms are:

$$\overline{H}_{12} = \tfrac{1}{2}(H_{12} + H_{21}) = \tfrac{1}{2}\{ \langle 1 | - \tfrac{1}{2}\nabla_1^2 + V_1 + V_2 | 2 \rangle$$
$$+ \langle 2 | - \tfrac{1}{2}\nabla_2^2 + V_1 + V_2 | 1 \rangle \} \qquad [35]$$

Now,

$$\langle 1 | - \tfrac{1}{2}\nabla_1^2 + V_1 | 1 \rangle = -I_1$$

$$\langle 1 | V_2 | 1 \rangle = R_{12}^{-1}$$

Furthermore, through Mulliken's approximation

$$\chi_1\chi_2 \simeq \frac{S_{12}}{2} (\chi_1{}^2 + \chi_2{}^2) \qquad [36]$$

and supposing this can be introduced into the off-diagonal core integral,

$$\langle 1 | -\tfrac{1}{2}\nabla_1{}^2 - V_1 | 2\rangle = \tfrac{1}{2}S_{12} \{\langle 1 | -\tfrac{1}{2}\nabla_1{}^2 + V_1 | 1\rangle$$
$$+ \langle 2 | -\nabla_1{}^2 + V_1 | 2\rangle \} \qquad [37]$$

Writing $\rho_1 = \langle 1 | V_1 | 1\rangle$, etc., we obtain

$$H_{12} = \frac{S_{12}}{2} \{ \tfrac{1}{2}(-I_1 - I_2) - 2R_{12}{}^{-1} - \tfrac{1}{2}(\rho_1 + \rho_2)\} \qquad [38]$$

This is our usual β, and Pohl's method is one of the rare semi-empirical methods where it is not considered an empirical parameter. Equation [37] is similar to the formula used in Hoffmann's EH calculations. For hydrogen 1s orbitals,

$$\rho_H = \langle 1\underline{s} | r^{-1} | 1\underline{s}\rangle = 1$$

For p_z orbitals,

$$\rho_\chi = \langle np_z | r^{-1} | np_z\rangle = \frac{\mu_n}{n^*},$$

where μ_n is the orbital exponent, and n* is the effective principal quantum number.

For one-center one-orbital repulsion terms, the Pariser-Parr IA formula is used. For two-center coulomb integrals, the point-charge approximation yields $R_{12}{}^{-1}$ and for mixed integrals we have:

$$(11|12) = \frac{S_{12}}{2} \{(11|11) + (11|22)\} = \frac{S}{2} \{I_1 + A_1 + R_{12}{}^{-1}\} \qquad [39]$$

and so forth.

There is a useful formula for core-core repulsions in (146). In a later paper Harris and Pohl (147) used split-shell molecular orbitals for σ electrons in hydrogen halides with different spatial orbitals for the two electrons.

There are many interesting applications in Pohl, Rein, and Harris' paper (146) relating to the calculation of bond energies, vibrational force constants, and bond dipoles.

A very important application of this method concerns hydrogen-bonded systems involving potential functions, tautomeric equilibria, and hydrogen bonding in biological systems. We can only refer here to papers by Rein and Ladik (148), Rein and Harris (149–152), MacIntyre and Löwdin (154) on the electronic energy of the DNA replication plane, and Rein and Svetina (155) on two-proton vibrational states in nucleic acid hydrogen bonds. (See Chapter 4.) Manne (153) outlined an interesting method in which many-center integrals are handled with Mulliken-type approximations, SCF atomic orbitals are used as a basis, and core-valence interactions are treated by perturbational methods.

The method of Yonezawa, Yamaguchi, and Kato (156) is another adaptation of Roothaan's LCAO SCF method. Overlap and differential overlap are retained. One-center one-orbital coulomb integrals are computed from the IA formula, the difference between the respective valence-state ionization potential and electron affinity. One-center exchange integrals were first neglected but later the authors discarded this approximation. Two-center coulomb integrals are computed from Ohno (132).

$$(rr \mid ss) = \frac{1}{\sqrt{a^2 + (R_{rs})^2}} \qquad [40]$$

where $1/a = (rr \mid rr) = I_r - A_r$. For the heteropolar case, I and A are taken to be the arithmetic means of the appropriate homopolar values for this purpose. Integrals containing differential overlap are taken care of by Mulliken's approximation. The "own atom" part of the diagonal core term is:

$$U_{rr} = -I_r - (N_r - 1)(rr \mid rr) - \sum_{r \neq r'} N_{r'}(rr \mid r'r') \qquad [41]$$

with

$$(rr \mid r'r') = \tfrac{1}{2}[(rr \mid rr) + (r'r' \mid r'r')] \qquad [42]$$

where N_r and $N_{r'}$ are the number of electrons occupying atomic orbitals r and r'. Attraction terms with other atomic cores are replaced by negatives of corresponding repulsion integrals.

$$(B \mid rr) = -Z_B(s_B \, s_B \mid rr) \qquad [43]$$

The Wolfsberg-Helmholtz formula is used for the off-diagonal core terms,

$$H_{rs} = 0.5 \, K \, S_{rs}(H_{rr} + H_{ss}) \qquad [44]$$

The results depend in a rather delicate way on the choice of K; in this treatment the value 1.1 is given. The starting coefficients for the SCF calculations are obtained by Hoffman's method. This is a reasonable scheme and it leads to encouraging results.

One-center exchange integrals were reintroduced in a later paper by Kato, Konishi, and Yonezawa (157). The formulas they used were

and

$$(sp \mid sp) = 0.045 \, Z_A \, (ss \mid pp)$$

$$(pp' \mid pp') = 0.011 \, Z_A \, (pp \mid p'p')$$

where Z_A now is the number of valence electrons on the respective atom. These formulas reproduce the values estimated by Hinze and Jaffé (158).

Introducing these and computing H_{rs} from a more refined formula:

$$H_{rs} = \frac{S_{rs}}{2} \left\{ -(Z_A + Z_B)/R_{AB} - (B \mid rr) - (A \mid ss) + (H_{rr} + H_{ss}) \right\} \qquad [45]$$

they obtained greatly improved results.

In subsequent papers Yonezawa and his collaborators treated the electronic structures of trans and cis butadiene, acrolein, and glyoxal (159), then turned their attention toward open-shell species in the manner of Pople and Nesbet (160), treating separately the spin orbitals with α and β spins. Their calculations were extended to a number of radicals and ionic intermediates (161, 162), in particular carbonium ions, alkyl cations, and protonated hydrocarbons (163). We refer the reader to these papers for further reference.

6. Methods using Slater-Condon parameters

Fischer-Hjalmars and Skancke worked out a method characterized by the use of Slater-Condon parameters for obtaining the one-center electron interaction integrals and by not making the zero differential overlap approximation. Fischer-Hjalmars (164) proposed a procedure of this type for π-electron problems which was subsequently adapted by Skancke (165, 166) to σ-electron systems.

In Slater's theory of atoms (167, 168), coulomb and exchange integrals can be expressed, apart from a spin factor, as:

$$J(n l m_l; n' l' m'_l) = \sum_k a^k (l m_l; l' m'_l) F^k(nl; n' l') \tag{46}$$

and

$$K(n l m_l; n' l' m'_l) = \sum_k b^k (l m_l; l' m'_l) G^k(nl; n' l') \tag{47}$$

where the a^k and b^k come from the angular parts of the wavefunctions and the F^k and G^k from their radial parts, and the summation is over the appropriate Legendre polynomials. The former can be computed and were tabulated by Condon and Shortley (168). The Slater-Condon parameters F^k and G^k can be computed or deducted from observed atomic excitation energies. Computed values can be found in an extensive work of Bingel (169) and empirical values in the paper of Pilcher and Skinner (20).

For 2 p electrons the one-center one-orbital coulomb integral is

$$(\mu\mu \mid \mu\mu) = F_0^{pp} + 4F_2^{pp} \tag{48}$$

Fischer-Hjalmars used Pilcher and Skinner's spectral values for F_2^{pp}. This cannot be done with F_0^{pp} since this term never appears as a difference between the energies of two atomic states. To circumvent this difficulty she supposed that the following equations hold:

$$\frac{F_2^{pp}}{(F_2^{pp})_t} = \frac{F_0^{pp}}{(F_0^{pp})_t} = \frac{(\mu\mu \mid \mu\mu)}{(\mu\mu \mid \mu\mu)_t} \tag{49}$$

where the index t refers to theoretical values computed with Slater orbitals. These theoretical values are in terms of the orbital exponent:

$$(F_0^{pp})_t = \frac{93}{256} \zeta$$

$$(F_2^{pp})_t = \frac{9}{1280} \zeta$$

and

$$(\mu\mu \mid \mu\mu)_t = \frac{501}{1280} \zeta \tag{50}$$

so that, assuming equation [49],

$$(\mu\mu \mid \mu\mu) = \frac{501}{9} F_2^{pp} \qquad [51]$$

Using the values which were given by Duncanson and Coulson (170), Fischer-Hjalmars obtained 16.61, 20.34, and 23.64 ev for C, N, and O respectively from equation [50]. On the other hand, from [51] using Pilcher and Skinner's experimental values, 11.90, 15.39, and 19.46 are obtained. The differences between the respective theoretical and empirical values are close to 5 ev. The near constancy of this value for first-row elements is in line with Clementi's (171) estimated correlation energies for these atoms; and the numerical value of the difference can be justified by some approximate arguments involving the virial theorem (164). Thus (though this may be fortuitous), correlation energy (defined as $E_{exact} - E_{HF}$) appears to be the main factor obliging us to scale down the theoretically computed values of $(\mu\mu \mid \mu\mu)$. The values finally adopted by Fischer-Hjalmars (164) were 11.76, 15.49, and 18.79 ev for C, N, and O in that order. (A more accurate treatment with explicit evaluation of valence-state HF and correlation terms giving the theoretical and possibly the most reliable values for C, N, and O, is found in Sinanoglu, Modern Quantum Chemistry, see (13).)

Fischer-Hjalmars' correction to the two-center coulomb integrals is also unconventional. The correction energy was first obtained in the case of the hydrogen molecule as the difference in energy obtained by a valence-bond wavefunction and Roothaan and Kolos' (172) exact wavefunction as a function of distance ("remainder energy"). (See Fig. 1 in (164).) The valence-bond wavefunction was chosen rather than the molecular orbital wavefunction because the former contains more correlation (173) and has the correct behavior at infinite distance. Then we may pass to p electrons by the transformation

$$(\mu\mu \mid \nu\nu)_t - (\mu\mu \mid \nu\nu) = \frac{N(\mu, \nu)^2}{(S_a, S_b)} f(\rho) \qquad [52]$$

where (μ, ν) and (S_a, S_b) are overlap integrals and the normalizing factor N is adjusted to obtain the limiting value for the correction to $(\mu\mu \mid \mu\mu)$ as obtained in the way described above. Polynomials in ρ can be fitted to [52] in order to facilitate calculations.

Skancke (165) adapted this procedure to σ-electron calculations. His method is a semiempirical version of Roothaan's LCAO SCF scheme where neither overlap integrals nor differential overlap are neglected.

The one-center two-electron integrals are treated in Fischer-Hjalmars' manner from the Slater-Condon parameters. If $x \equiv 2p_x$, we have that

$$(xx \mid xx) = F_0^{pp} + 4F_2^{pp} \tag{53}$$

$$(xx \mid yy) = (xx \mid xx) - 6F_2^{pp} \tag{54}$$

and

$$(xy \mid xy) = 3F_2^{pp} \tag{55}$$

The integral $(ss \mid ss)$ where $s \equiv 2\underline{s}$ is obtained by the assumption that

$$\frac{F_0^{pp}}{F_0^{pp}\,'t} = \frac{F_0^{ss}}{F_0^{ss}\,'t} = \frac{(ss \mid ss)}{(ss \mid ss)_t} \tag{56}$$

where the index t again stands for theoretical.

There remain $(ss \mid xx)$ which is equal to $(ss \mid ss)$ for Slater orbitals and $(sx \mid sx) = G_1(s,p)$, where G_1 is given by Pilcher and Skinner (20).

Two-center two-electron integrals were again treated in Fischer-Hjalmars' manner as described above. For hybrid integrals a slightly modified form of Mulliken's approximation was used.

The one-electron integrals were adjusted to obtain agreement with the first ionization potential and first electronic excitation energy ($\sigma \rightarrow \sigma*$) of the ethane molecule where the orbitals on the carbon atoms were the only ones considered explicitly.

The two-center core integral β was expressed as

$$\beta = kS\alpha$$

following a procedure due to Mulliken (18). The best results were obtained with $k = 1.2$.

To determine the one-center core integral α, Skancke used Field and Franklin's (174) vertical ionization potential of ethane

(11.65 ev) which he set equal to $\alpha + \beta + J_{00}$ (175). It is interesting that using the value of α obtained this way and introducing the Goeppert-Mayer and Sklar (176) approximation, one obtains -14.77 ev for the valence-state ionization potential of a carbon atom in ethane which is very close to its value deduced by Pilcher and Skinner (20) from atomic spectral data.

Skancke applied his method to propane considered as a four C–C σ-electron problem with encouraging results.

We should like to point out that it is not definitely proven that the first ionization of the ethane molecule leaves the hole in the C–C bond; nor was the spectrum of ethane measured with sufficient accuracy when Skancke published his paper. Furthermore, the placing of the CH electrons into the core is an arguable procedure. All this, however, could affect only the numerical values of the parameters and the complexity of the calculations, and does not detract from the value of this interesting scheme.

Fisk and Fraga (177) computed Slater-Condon parameters from analytical HF functions for all atoms and for their positive and negative ions from He to Zn. They do not seem to have been used in semiempirical calculations so far.

7. New lines of progress

The foundations of the various approximate methods which were applied to σ-electron problems were critically examined by Cook, Hollis, and McWeeny (178).

Instead of ignoring all but the largest integrals, these authors use "a particular basis of orthogonalized orbitals chosen so as to render valid, with a fair degree of accuracy, all the integral approximations subsequently made." This then allows them to "simulate" a nonempirical calculation based upon a carefully chosen orbital set. Actually the way in which the π-electron Pariser and Parr (86) ZDO approach is justified, is to consider it based on a set of symmetrically orthogonalized 2pπ orbitals in Löwdin's manner (58). However, as Cook et al. (178) point out, "when there are several AO's on each center in the molecule, the symmetrical orthogonalization procedure is not sufficient to ensure the validity of a ZDO approximation." Some other way has to be found to construct a basis which can achieve this.

Furthermore, when the coulomb integrals are averaged as in the CNDO method, invariance with respect to the choice of local axes is only restored in a mathematical sense; it amounts to a change of basis and there is no insurance that approximations made in one basis remain valid in the other. Actually non-coulombic two-electron integrals are negligible only when the basis orbitals are orthogonal and localized mainly in different regions of space.

Thus the integral approximations are linked to the choice of basis. Cook et al. found that an appropriate basis consists of orthogonal orbitals possessing a maximum of localization. The $\underline{s}$, $\underline{p}$, $\underline{d}$, . . . orbitals at a given center are of course orthogonal but they are not highly localized or separated from each other ($2\underline{s}$ and $2\underline{p}$, for example). Hybrid orbitals have much higher localization. Hybrid orbitals on different centers are far from being mutually orthogonal but they can be symmetrically orthogonalized by Löwdin's procedure. Such orthogonalized hybrids were used previously by Klessinger and McWeeny (179) and by McWeeny and Ohno (180).

On the other hand, inner-shell orbitals should not be mixed with valence orbitals but the latter should be Schmidt orthogonalized against the former before constructing the hybrids. Integrals based on localized orthonormal orbitals are often transferable from one molecule to the other.

Cook, Hollis, and McWeeny performed nonempirical calculations on H_2O and CH_4 using the orthogonal hybrid basis. They did this in both the SCF MO scheme and the SCGF (self-consistent group function) scheme (181, 179) in which the wavefunction contains one factor for "each recognizable group of electrons" (bond, lone pair, inner shell). One-electron integrals were computed exactly. Two-electron integrals were also computed exactly over the orthogonal hybrid basis and at first were all included; then in a second calculation, the small two-electron integrals which would be omitted in the CNDO approximation were omitted; and, in a third calculation, the one-center exchange integrals were retained as in the NDDO approximation.

The CNDO calculation did not reproduce well the results of the full calculation but the NDDO calculations did for both electron distribution and bond polarities.

The authors conclude that only the NDDO scheme can be used to simulate the results of a nonempirical calculation based on symmetrically orthogonalized hybrid orbitals and that this should be the starting point for semiempirical calculations. (The INDO method (108, 109) was initiated after (178) was published.)

The prescription is that two-electron integrals should be
calculated first over nonorthogonal hybrid atomic orbitals and
then those of the one-center coulomb type should be increased
by about 12% and the two-center coulomb integrals decreased
by 12%. One-center exchange integrals do not need adjustment
and other types of integrals can be neglected. The one-electron
Hamiltonian matrix is accurately computed on the orthogonal
hybrid basis.

The authors tested this scheme on ethane, ethylene, and
acetylene with very encouraging results. The SCGF framework
presents some significant computational advantages (a much
lower number of β's in particular). Some of their results are

Table 6. Orbital populations for water and methane

Water

		h_1	b_1	l_1
	CNDO	1.10	0.95	1.95
SCF MO	NDDO	0.89	1.15	1.95
	All integrals retained	0.88	1.12	2.00
	CNDO	1.04	0.96	2.00
SCGF	NDDO	0.88	1.12	2.00
	All integrals retained	0.85	1.15	2.00

Methane

		h_1	b_1
	CNDO	1.10	0.90
SCF MO	NDDO	0.97	1.03
	All integrals retained	0.92	1.08
	CNDO	1.06	0.94
SCGF	NDDO	0.95	1.05
	All integrals retained	0.92	1.08

After Cook, Hollis, and McWeeny (178).

presented in Table 6, where h stands for hydrogen $1\underline{s}$, b for bond hybrid pointing toward hydrogen 1, and l, for a lone-pair hybrid orbital. Exact (nonempirical) integrals were used.

The results shown in Table 7 were obtained as described above, that is, by calculating the one-electron Hamiltonian matrix and estimating the NDDO two-electron integrals, both on the orthogonal hybrid basis. We give only the SCGF values. One interesting feature is the increasing polarity of the C–H bond in the order methane < ethane < ethylene < acetylene.

Table 7. Orbital populations for methane, ethane, ethylene, and acetylene

Orbital populations

	h_1	b_1	σ_1	π_1	Energy (Hartrees)
Methane	0.92	1.08	-		−52.84
Ethane	0.87	1.13	1.00		−119.85
Ethylene	0.82	1.18	1.00	1.00	−110.42
Acetylene	0.76	1.24	1.00	1.00	−100.77

After Cook, Hollis, and McWeeny (178).

There is some interest in comparing the values of the integrals computed by Cook et al. over nonorthogonal hybrids and orthogonal hybrids respectively. These values are presented in Table 8 for the water molecule, all in atomic units ($h_1 \equiv H1\underline{s}$, $b_1 \equiv O\underline{sp}^3$ in bond 1, $l_1 \equiv O\underline{sp}^3$ lone pair). The main observation is the dramatic difference in the values of the two-center integrals containing differential overlap. They are clearly negligible in the orthogonal basis only.

Table 9 shows some of their core integrals, again in atomic units. The two-center integrals change almost by an order of magnitude. The large values of the α are due to the large one-center kinetic and coulomb integrals of which there are a few and not just one as in π-electron problems.

The selection of hybrid orbitals adequate to form a basis for the calculations is not always possible from chemical evidence or tradition. Obvious examples are "bent bonds," lone pairs, or the CO molecule. This problem has been treated

Table 8. Two electron integrals for the water molecule

	Over nonorthogonal hybrids	Over orthogonal hybrids
$(h_1h_1 \mid h_1h_1)$	0.625	0.652
$(b_1b_1 \mid h_1h_1)$	0.591	0.500
$(l_1l_1 \mid h_1h_1)$	0.441	0.388
$(l_1b_1 \mid l_1b_1)$	0.073	0.079
$(l_2l_2 \mid l_1l_1)$	0.750	0.751
$(h_1b_1 \mid h_1b_1)$	0.251	0.014
$(h_1h_1 \mid h_1b_1)$	0.361	0.025
$(b_1b_1 \mid h_1b_1)$	0.458	0.023

After Cook, Hollis, and McWeeny (178).

Table 9. Core integrals, in atomic units, for water

	α			β						
	h_1	b_1	l_1	$h_1{-}b_1$	$h_1{-}b_2$	$h_1{-}l_1$	$b_1{-}b_2$	$b_1{-}l_1$	$l_1{-}l_2$	$h_1{-}h_2$
Non-orthogonal hybrid	-3.92	-5.52	-5.49	-2.89	-0.92	-1.00	-0.08	-0.13	-0.20	-1.74
Orthogonal hybrid	-3.20	-5.58	-5.37	-0.46	-0.16	-0.17	-0.15	-0.13	-0.08	-0.06

After Cook, Hollis, and McWeeny (178).

extensively by McWeeny and Del Re (182) who described the "best possible" choice in terms of localized orbitals constructed by combining suitable hybrids on adjacent atoms. This problem is beyond the scope of the present review, however.

Berthier, Del Re, and Veillard (183) examined some frequently used approximations for the off-diagonal elements in semiempirical calculations, such as the proportionality of β to the overlap integrals and the Wolfsberg-Helmholtz formula. They pointed out that to be reliable these must fulfill the requirement of co-

variance with respect to linear transformations of the basis and
to shifts in the zero point of the energy. If they do not, the basis
and the zero point must be specified.

$$\beta = k \, (S - I) \tag{57}$$

is not, in general, covariant with respect to linear transformations. It is covariant for changes in the zero point of the energy,
if k is a function of the zero point.

$$\beta_{pq} = k \, S_{pq} \, (\alpha_p + \alpha_q)/2 \tag{58}$$

is not covariant under linear transformations of the basis and is
invariant for changes in the zero point of energy only if $k = 1$.
Interpretation of the results obtained through these formulas is
given in (183).

Carpentieri, Porro, and Del Re (185) subjected to critical
examination the formulas of Del Re and Parr (184). These are
expressions for the energy matrix elements in the LCAO MO
method including configuration interaction. Their paper concerns π-electron systems but many of their conclusions are ap-
plicable to σ-electron systems as well.

Ciampi and Paoloni (186) worked out a semiempirical method based on geminal functions and applied it with Hückel molecular orbitals to π and n electrons. This recent work may be the
starting point for further interesting developments.

Introducing correlation explicitly into semiempirical methods dealing with all-valence electron problems is the next task
for quantum chemists, and it will certainly lead to many new
developments. This field has been recently opened by Sinanoğlu
and Skutnik (187) and by Trindle and Sinanoğlu (188, 189). These
authors found that it is advantageous to convert the molecular
(symmetry) orbitals into localized orbitals. This was done a
long time ago in the framework of the simple Hückel method by
Hall and Lennard-Jones (190, 191) and by Lennard-Jones and
Pople (192) (self-energy localization procedure). Edmiston and
Ruedenberg (193—95) investigated the nature of localized atomic
and molecular orbitals in more elaborate methods. Trindle and
Sinanoğlu had done this in semiempirical methods (CNDO/2).
They showed how to obtain localized orbitals which are chemically significant and can be transferred from molecule to molecule
in certain cases. Hybridization was also examined by them in
this connection. Sinanoğlu and Skutnik have further shown that

correlations between neighboring tetrahedral orbitals contribu
about twice as much to the valence-shell correlation energies,
in the case of neon and methane, as the correlations within the
four doubly occupied tetrahedral pairs. This is a surprising r
sult and is likely to find general application.

Pamuk and Sinanoğlu (196) signalled a semiempirical version of Sinanoğlu's Many-Electron Theory based on their local
ized orbitals in which correlation would receive adequate representation.

8. Attempts to treat electronic spectra

Concerning electronic transitions and excited states, quantum
chemistry is still in its infancy. In the most popular closed-
shell Roothaan SCF method, only the ground-state wave functior
is made self-consistent. To consider the excited states to the
same degree of accuracy, open-shell SCF procedures should be
used, but this has seldom been done. Furthermore, it is questic
able if the same basis functions that are relatively good for the
ground state are as good for the excited states. We do not know
how to change the Z numbers in going from one state to the othe
(197). Correlation is likely to have a different effect in differen
states (198). Higher atomic orbitals should probably be include
into the basis. Valence-shell and Rydberg transitions are expected to mingle a great deal in σ-electron spectra. Probably
the best tool for dealing with electronic spectra is configuration
interaction which takes partial care of correlation and does not
favor the ground state one-sidedly.

Only a few schemes have been offered for the treatment of
the electronic spectra of σ electrons.

Katagiri and Sandorfy (199) presented a Pariser and Parr
type method, based on Csp^3 hybrid and $H1\underline{s}$ orbitals. They applied it to methane, ethane, and propane. Carbon one-center
coulomb integrals were determined in Fischer-Hjalmars' manner (164) through the Slater-Condon parameters. In Bingel's
work (169) they are given as functions of the orbital exponent
$\zeta = Z/2$. Since, on the other hand, the F^k and G^k can be determined from observed atomic spectra (see Pilcher and Skinner
(20)), we have a means for deducing empirical Z values for given
cases. The reduced values are as follows:

	C^+	C	C^-
C2p	2.42	2.23	1.89
C2s	2.45	2.26	1.92

2.23 compares to Slater's 3.25, a drastic change. The integrals which were used were theoretical integrals computed with these Z numbers and the C^- values for the one-center coulomb integral.

For hydrogen this method was not used because of lack of spectral data on H^-. The (hh | hh) was estimated to be 11.542 ev, using Pekeris' (200) electron-affinity value and the Goeppert-Mayer and Sklar (176) approximation. The theoretical value, with Z = 0.69, is 11.734 ev. Integrals containing differential overlap were neglected except for one-center exchange integrals.

Two-center coulomb integrals for internuclear distances larger than 2.8Å were computed from the uniformly charged sphere model (86, 88) for carbon 2p orbitals and by the point-charge model for C 2s and H 1s orbitals. Pariser and Parr's formula $(ar + br^2)$ was used for shorter distances (88).

Mulliken's definition (18), $\beta = \gamma + S\alpha$, was used for the core integrals where α and γ were explicitly written out using the Goeppert-Mayer and Sklar approximation. Then Mulliken's approximation was applied to differential overlap leaving finally,

$$\beta_{pq} = \frac{-S_{pq}}{2} \left[C_{pq} + \tfrac{1}{2}(pp \mid pp) + \tfrac{1}{2}(qq \mid qq) - (pp \mid qq) \right]$$

[59]

where

$$C_{pq} = \tfrac{1}{2}[(p:pp) + (q:qq)]$$

[60]

(Flurry and his co-workers (201) used this in their PPP-type calculations of heteroatomic systems.) C_{pq}, which contains the one-center penetration integrals, was kept as a parameter and was adjusted to fit the first singlet-singlet transition energy of methane. The β_{tt}, between two hybrids on the same carbon was also treated as a parameter. Its value was -1.32 ev, for the results quoted below.

The first ionization potentials turned out to be too high by about 2 ev, a familiar situation in π-electron calculations. The first electronic transition is at 10.26 ev ($^1F_2 \leftarrow {}^1A_1$) in methane, at 7.81 and 7.84 in staggered and eclipsed ethane respectively ($^1E \leftarrow {}^1A$), and at 7.53 in all trans propane. These results inter-

pret well the observed bathochromic trend with increasing chain-
length in normal paraffins (202, 203), and the predicted shifts are
of the right order of magnitude. The assignments are in agreement with Mulliken's united-atom treatment (204, 205). Energy
differences between rotational isomers are of the order of tenths
of ev's. Interestingly, the first ionization seems to leave the hole
in the C–H rather than in the C–C bonds for the shortest paraffins. This is a result contrary to tradition but compatible with
photoelectron spectra measured by Al-Joboury and Turner (206).

The charge densities on the atoms are:

	H	C
methane	+0.059	−0.236
ethane	+0.053	−0.160

which are certainly of the right order of magnitude (see (178))
and much smaller than Hoffmann's (22) (if no iterations are applied to the latter).

Katagiri and Sandorfy (199) did not apply configuration interaction to their method. This has been done, however, by Bessis
(207) with but little change in the results. This is likely to be a
matter of chance.

In an interesting paper, Brown and Krishna (208) treated
the case of cyclopropane. Their treatment is of the Pariser and
Parr type in which only the six σ orbitals providing the C–C
bonds are treated explicitly. They assumed that the C–H bond
would be little affected in the transitions of lowest energy. This
assumption may be more justified in the case of a quasi-con-
jugated molecule like cyclopropane than for other paraffins, but
it is, of course, still questionable.

Brown and Krishna chose their atomic orbital basis the fol-
lowing way. Three sp^2 orbitals χ_1, χ_2 and χ_3 are directed so
that their symmetry axes intersect at the center of the cyclopropane triangle; three 2p atomic orbitals, χ_a, χ_b, χ_c, are ori-
ented in such a way that their nodal planes intersect along the
threefold symmetry axis of the molecule. The simple Hückel
treatment yields a scheme of energy levels very similar to that
of benzene (in the π approximation). The Hückel orbitals are
then used as the starting point of the Pariser and Parr treatment. For obtaining the one-center core integrals the Goeppert-
Mayer and Sklar approximation and valence-state ionization

potentials were used. As to the resonance integrals, Brown and Krishna argue that allowance should be made for the fact that in homonuclear diatomic molecules it is directly proportional to the core charge on the nuclei. "The same dependence on core charge should apply for a more extended molecule in view of the localized property of the resonance integral." They therefore used the formula

$$\beta_{\mu\nu} = (\chi_\mu \chi_\nu)^{1/2} \beta^\pi_{CC} S_{\mu\nu}/S^\pi_{CC} \qquad [61]$$

where χ_μ is the core charge for μ (+2) and the proportionality between resonance and overlap integrals is assumed. β^π_{CC} was chosen to fit observed π excitation energies by Pariser and Parr,

$$\beta^\pi_{CC} = -2517.5 \exp(-5.0007R)\underline{ev} \text{ where R is in}$$
$$\text{angstroms.}$$

One-center coulomb integrals (γ) were computed on the basis of the assumption that

$$\gamma^{Corr}_{2\underline{s}, 2\underline{s}} \Big/ \gamma^{Slater}_{2\underline{s}, 2\underline{s}} = \gamma^{Corr}_{2\underline{p}, 2\underline{p}} \Big/ \gamma^{Slater}_{2\underline{p}, 2\underline{p}} \qquad [62]$$

where the γ^{Slater} are evaluated theoretically using Slater functions. The $\gamma^{Corr}_{2\underline{p}, 2\underline{p}}$ is obtained from Pariser and Parr's IA or Paoloni's (89)

$$\gamma = 3.29 \text{ Z } \underline{ev} \qquad [63]$$

which implies a correction factor of $k_1 = 0.62$ in

$$\gamma = k_1 \gamma^{Slater}$$

Brown and Krishna used the same factor for one-center two-orbital coulomb integrals.

Differential overlap was neglected, except if in (ab| cd) a and b and c and d are orbitals on the same atom. For two-center coulomb integrals the same correction factor 0.62 was used on the grounds that the correlation correction should be comparable to the one for one-center integrals because of the large overlap of σ electrons. This is an interesting idea capable of further refinements.

Brown and Krishna applied configuration interaction between the doubly excited levels and estimated the probable effect of introducing more configuration interaction.

One $^1A'_2$ and two $^1E'$ excited states are predicted whose order depends on the amount of configuration interaction applied. The lowest frequency part of the spectrum would consist of these three bands and one singlet-triplet band to a $^3A'_2$ state. This is a good basis for discussion of the electronic spectrum of cyclopropane.

Imamura and his co-workers (209) applied a semiempirical SCF treatment to the peptide link with both σ and π electrons included explicitly. The molecular orbitals were built from four $2p\pi$ orbitals, including the hydrogen $2p\pi$ orbital, and from six σ orbitals.

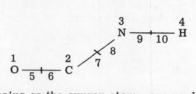

The lone pairs on the oxygen atom were neglected. The σ and π systems were treated separately. The σ problem was treated on the basis of hybrid sp^2 orbitals which were used to carry out a Pariser and Parr calculation. One-center core integrals were handled through the Goeppert-Mayer and Sklar approximation; the β were set proportional to the overlap integrals. The β between two hybrid orbitals on the same atom were estimated from the formula

$$(\underline{sp}_1{}^2 \,|\, H \,|\, \underline{sp}_2{}^2) = \tfrac{1}{3}\left\{(2\underline{s} \,|\, H \,|\, 2\underline{s}) - (2\underline{p} \,|\, H \,|\, 2\underline{p})\right\} \qquad [64]$$

where $(2\underline{s} \,|\, H \,|\, 2\underline{s})$ and $(2\underline{p} \,|\, H \,|\, 2\underline{s})$ were replaced by the respective ionization potentials. For the coulomb integrals, IA and the Mataga-Nishimoto (91) formula were used. Differential overlap was neglected. Valence-state ionization potentials of Hinze and Jaffé (21) were used.

Figure 5 shows the sequence of orbitals found by Imamura et al. (with 0.3625 as the effective nuclear charge for the hydrogen $2p\pi$ orbital). The spectral predictions were as shown in Table 10.

It is thus predicted that there are weak π-σ* and σ-π*transitions at frequencies lower than the first strong observed π-π* band at 7.22 ev. The next strong π-π* band is at 9.22 ev and its intensity turns out to be much too low in the calculations.

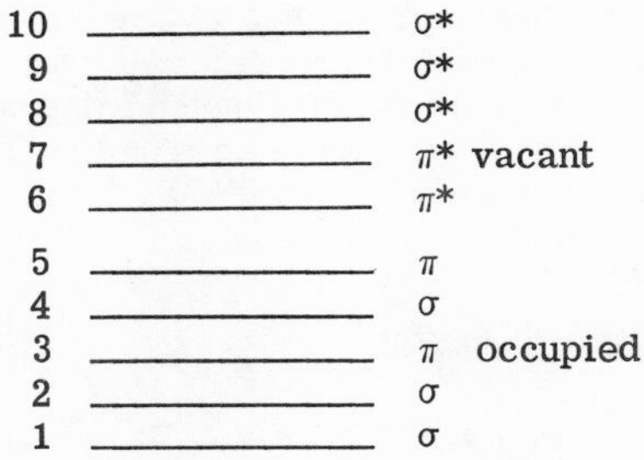

Fig. 5. Sequence of orbitals for the peptide link (Imamura et al. (209)).

Table 10. Excitation energies of the peptide link (in $\underline{ev}$)

Transition	Excitation energy	Type of transition	Oscillator strength
5 → 8	4.50	π-σ*	Very small
5 → 6	6.00	π-π*	0.002
4 → 7	6.26	σ-π*	Very small
5 → 9	6.97	π-σ*	Very small
5 → 7	7.04	π-π*	0.455
3 → 8	7.65	π-σ*	Very small
4 → 8	9.03	σ-σ*	
3 → 6	9.23	π-π*	0.002
4 → 6	9.27	σ-π*	Very small
3 → 7	9.79	π-π*	0.126

After Imamura et al. (209).

The effect of the hydrogen bonds on the semiconductivity of proteins is also discussed.

Del Bene and Jaffé (210) applied the CNDO method to benzene, pyridine, and the diazines with the aim of determining the influence of the σ electrons on n and π electron levels. To do this they completed the CNDO method with first-order configuration interaction and wrote a program for it. (The 30 lowest singly excited states were included.)

In order to obtain excitation energies in fair agreement with experiment, Del Bene and Jaffé had to change two para-

meters in the CNDO scheme. They used IA for one-center coulomb integrals and Pariser and Parr's formulas for the two-
center coulomb integrals. For the resonance integrals they
used the Pople et al. (98) formula but with a parameter K dif-
ferentiating between the σ and π cases:

$$\beta_{\mu\nu}{}^{\sigma} = \tfrac{1}{2}(\beta_A{}^O + \beta_B{}^{O2})S_{\mu\nu}$$

$$\beta_{\mu\nu}{}^{\pi} = \tfrac{1}{2}K(\beta_A{}^O + \beta_B{}^O)S_{\mu\nu}$$

The best value of K turned out to be 0.585. These changes
were rendered necessary because the atom averaging of these
integrals caused excessive mixing of σ and π orbitals. Also the
$\beta_A{}^O$ had to be readjusted.

With this modified CNDO scheme they obtained very good
predictions for the (singlet-singlet) n $\rightarrow$ π^* and first three
$\pi \rightarrow \pi^*$ transitions, as shown in Table 11. Their second excitation energy is invariably low and they discuss the causes of this.

Table 11. Excitation energies for benzene, pyridine,
and the diazines (in ev)

	n $\rightarrow$ π^*	$\pi \rightarrow \pi^*$	$\pi \rightarrow \pi^*$	$\mu \rightarrow \mu^*$
Benzene				
obs.		4.7	6.1	6.9
calc.		4.7	5.2	6.9
Pyridine				
obs.	4.3	4.8	6.2	7.0
calc.	4.3	4.8	5.4	7.1
1-2 Diazine				
obs.	3.3	4.9	6.2	7.1
calc.	3.4	4.9	5.4	7.1
1-3 Diazine				
obs.	3.9	5.0	6.5	7.3
calc.	4.2	5.0	5.6	7.3
1-4 Diazine				
obs.	3.8	4.8	6.3	7.5
calc.	3.6	4.7	5.4	7.3

After Del Bene and Jaffé (210).

An interesting result of these calculations was that the n electrons are significantly delocalized throughout the other σ orbitals. (See Brown and Harcourt (90).) In pyridine the energy of the lone pair is immediately below the two highest occupied π orbitals of the molecule. This is in agreement with the fact that the first ionization potential leaves the hole in a π-molecular orbital (211, 212). The s character of the lone-pair orbital turns out to be only 14%.

In two subsequent papers (213, 214) Del Bene and Jaffé extended their work to five-membered ring molecules (cyclopentadiene, pyrrole, furan, pyrazole, imidazole, etc.) and to monosubstituted benzenes and pyridines. Their calculations give support for the hyperconjugative effect of the CH_2 group in the case of cyclopentadiene and indicate that the third weak electronic transition (from the low-frequency end) is of $\sigma \rightarrow \pi^*$ character (the order is 4.8 ev ($^1B_2$), 6.2 ($^1A_1$), 7.5 ($^1B_1$), and 7.9 ($^1A_1$)), the three other bands being $\pi \rightarrow \pi^*$. All four transitions are $\pi \rightarrow \pi^*$ in pyrrole and furan. In furan the n $\rightarrow \pi^*$ transition would be at very high frequencies, the n level being very low.

The method is very successful in interpreting substitution effects for both benzene and pyridine derivatives. Charge transfer seems to be unimportant for the first four $\pi \rightarrow \pi^*$ transitions.

Kroto and Santry (215) approximated the excited-state wavefunctions by a single determinant constructed using the virtual orbitals obtained from the ground-state iteration procedure in the CNDO scheme. They found the excitation energies generally too high but obtained useful results relating to molecular geometry in excited states.

There have been further attempts to assess the effect of the σ electrons on the $\pi \rightarrow \pi^*$ spectra. Clark and Ragle (216) treated benzene and ethylene. They also used the CNDO method but had to change certain parameters. We reproduce here two of their tables, Tables 12 and 13, in which they compare their computed excitation energies to their experimental values and to the results of Moskowitz and Harrison (217) and Shulman and Moskowitz (218) who made SCF calculations using basis sets of Gaussian orbitals.

Song and Moore (219) and Song (220) made calculations on the spectra of halogenated purines and pyrimidines, and Allinger et al. (221) on cyclobutadiene and dimethyl-cyclobutadiene taking some account of the σ electrons. The influence of the σ system on the $\pi \rightarrow \pi^*$ transitions of linear polyenes was studied by Denis and Malrieu (222).

Table 12. Estimated transition energies of benzene (in ev)

Term symmetry	Experimental energy	Shulman and Moskowitz		Present work
$^3B_{1u}$	3.66	5.11	4.41	3.66
$^1B_{1u}$	6.09	7.37	7.24	5.88
$^1B_{2u}$	4.72	6.91	6.43	5.50
$^1E_{1u}$	6.82–6.93	10.1	9.89	7.63 10.72
$^3B_{2u}$		6.35	6.04	5.50
$^3E_{1u}$		5.46	5.23	8.18 4.58

After Clark and Ragle (216).

Table 13. Excitation energies of planar ethylene (in ev)

Term symmetry	Experimental energy	Moskowitz and Harrison	Present work
B	3.56	4.32	3.1
B	4.96	10.43	5.8
B			7.9
A		15.28	11.2

After Clark and Ragle (216).

The program of Del Bene and Jaffé (210) was used by Sala-hub and Sandorfy (223) to compute the σ energy levels for a few normal and branched chain paraffins. They found fair agreement with experiment.

Kato and co-workers (224) carried out extended Hückel calculations including 3s and 3p orbitals for carbon and 2s and 2p orbitals for hydrogen. They found a very significant lowering of the energies of the vacant MO. This is an important result for the treatment of electronic spectra and excited states.

All this is still in a stage of exploration. More rigorous studies are plainly needed. A recent work of Skutnik et al. (225)

on correlation effects in the excited states of atoms, and the calculation of Rydberg orbitals by several authors should turn out to be helpful. A number of references for the latter are found in a paper by Hosoya (226) who computed Rydberg orbitals for the atoms Li to F.

One of the most significant developments in the understanding of the spectra of organic molecules was Berry's (227) assignment of the medium intensity band, which appears at the low-frequency side of the strong $\pi \rightarrow \pi^*$ band in the spectra of olefins, to a $\sigma \rightarrow \pi^*$ transition. Robin et al. (228) showed that the transition is actually $\pi \rightarrow \sigma^*$, the excited σ^* orbital having appreciable Rydberg character. They substantiated this assignment by extending the basis of the Gaussian calculations of Moskowitz and Harrison (217) to include higher orbitals and by experimental arguments. The existence of low-lying σ^* Rydberg levels obliges us to rethink many of our conventional ideas on the electronic spectra of organic molecules.*

*Recently Merer and Mulliken (Can. J. Phys. 47, 1731 (1969)) and McDiarmid (J. Chem. Phys. 50, 1794 (1969)) have shown that this transition is best described as the first member of a Rydberg series shifted to lower frequencies from its position in ethylene.

Chapter 3

Methods Based on Bond Orbitals and Polyelectronic Functions

1. <u>Localizability of electrons</u>

The behavior of some atomic and molecular properties suggests the existence of regions within atoms and molecules in which electrons are localized. It is well known, for example, that magnetic rotation (229), magnetic susceptibility (230), and dipole moment can be empirically divided into components associated with bonds, lone pairs, and atomic cores.

However, the indistinguishability of the electrons makes it difficult to localize them. As a result, the wavefunction possesses symmetry properties. It is antisymmetric with respect to any permutation of the space and spin coordinates associated with two electrons. As a consequence, the mean value of any property A_i associated with a given electron

$$\bar{A} = <\psi \mid \hat{A}_i \mid \psi>$$

where $\hat{A}$ denotes the operator associated with A, is the same for all the electrons of an atom or a molecule. For example, the average of the distance of an electron to the nucleus of an atom is the same for all the electrons of this atom. Therefore, from the wave-mechanical viewpoint it is not convenient to distinguish between core electrons, valence electrons, bond electrons, etc.

Furthermore, it is customary but not correct to speak of K electrons, L electrons, σ electrons and π electrons. It would be much more satisfactory to speak only of K, L, σ, or π orbitals. To elaborate on this statement, let us consider a helium atom in a triplet state approximated by the functions:

$$\Psi(\overset{1}{M_1}, \overset{2}{M_2}) = \frac{1}{\sqrt{2}} \left[\psi_K(M_1)\,\psi_L(M_2) - \psi_K(M_2)\,\psi_L(M_1)\right]^*$$

Let us calculate the probability dp of finding electron 1 in a volume dv_1 near the point M_1. This is given by the expression:

*1 and 2 above M_1 and M_2 recall that the first point in the argument of a wavefunction refers to the point where we look for electron 1 and the second to the point where we wait for electron 2.

$$dp = dv_1 \int |\Psi(M_1, M_2)|^2 dv_2$$
$$= \tfrac{1}{2}[|\psi_K(M_1)|^2 + |\psi_L(M_1)|^2]$$

Clearly the value of this probability, and therefore the "motion" of electron one, depends on both orbital ψ_K and orbital ψ_L and not only on one of them. There is not a simple direct relationship between a given electron and a given orbital.

But if it is not possible to distinguish between the electrons, it is possible to designate regions in an atom or a molecule in which there are high probabilities of finding a given number of electrons (231, 232). Let us go back to our helium atom in its first excited state, but now let us represent it with a very elaborate wavefunction such as those proposed by Hylleraas. Let us consider a sphere of radius r (this value being completely arbitrary) with its center at the nucleus. It is possible to calculate the probability P of finding one electron, and one only, in this sphere. When r is very small this probability is also very small because the sphere is generally empty. When r is very large, P again will be very small because now the sphere will generally contain two electrons (and not one only). Thus, intuitively, we must anticipate that P will possess a maximum for at least one value of r. The curve of Figure 6 (233) shows that this is true. The maximum is large as it corresponds to P = 0.93. The corresponding radius is 1.7 a_o (a_o = 0.529Å being the atomic unit of length). We shall say that the best division of the atomic space into spherical loges is obtained when r = 1.7 a_o and that there is a probability of 93% of finding one electron, and only one, in this sphere (the other one being outside).

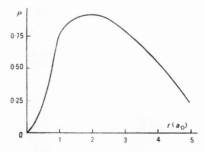

Fig. 6. The probability P as a function of r (Daudel et al. (233)).

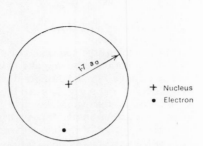

Fig. 7. The best decomposition in loges for the helium atom (first excited state) (Daudel et al. (233)).

Therefore Figure 7 symbolizes the most probable organization of the electrons of helium in its first excited state. We are thus quite naturally led to associate the sphere of radius 1.7 a_0 with the K shell and the rest of the space with the L shell. It is important to point out that the K and L shells are associated with some portion of space, but not with a particular electron.

These results can be extended. The space associated with an atom can be cut up into spherical rings, all concentric with the nucleus, one built on the other in such a way that there is a high probability of finding in each ring a certain number of electrons (234). For example, in the fluoride ion F^- (ground state) there is an 81% probability of finding two electrons of opposite spin in a sphere with the center at the nucleus and with radius $r = 0.35$ a_0, the other 8 electrons being outside. The sphere corresponds to the K loge, the remaining part of the space corresponding to the L loge (Fig. 8).

If the volume of a given loge is divided by the number of electrons which it usually contains, a certain volume v is obtained which gives an idea of the space associated with one electron in the loge. Moreover, we can evaluate the average value p of the electronic potential which is exerted in the loge. Odiot and Daudel (235) observed that for all atoms and all shells the following relation applies:

$$p^{3/2}v = \text{constant}$$

A kind of Boyle-Mariotte law exists between the "electronic pressure" p and the volume v associated with one electron in an atomic loge.

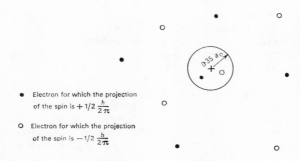

Fig. 8. The best decomposition in loges for the fluorine negative ion (Daudel et al. (233)).

The concept of loge is also helpful in describing the electronic structure of molecules. In a good division of a molecule into loges, some loges usually appear which were also representative of the free atoms before bonding. Such loges are said to be loges of the cores, while the others can be called loges of the bonds. As an example let us take the case of the lithium molecule Li_2. Figure 9 (233) represents a good division of this molecule into

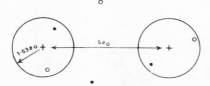

Fig. 9. A good decomposition in loges for the lithium molecule (Daudel et al. (233)).

loges. In the two spherical loges, which look like the K loges of the lithium atoms, there is a probability of 0.96 of finding two electrons (and two only) with opposite spins. Therefore the distribution of the electrons symbolized by Figure 9 has a very high probability. The region of space outside of the two spheres where there is also a high probability of finding a pair of electrons, can be considered as the loge corresponding to a two-electron bond.

This way of thinking can be used with any kind of molecules. It can be shown that between the cores of small atoms near the bond axis there is only room for two electrons with opposite spins. Then, for a molecule in which the number of electrons commonly found outside of the core regions is twice the number of adjacent atomic cores, a good division in loges consists of regions extended between two such adjacent cores in which there is a high probability of finding two electrons with opposite spins. Such loges correspond to localized two-electron bonds. Following this idea, the space associated with a methane molecule can be divided into a carbon K core region, in which there is a high probability of finding two electrons with opposite spins, and four C—H bond loges, in which there is also a high probability of finding a pair of such electrons.

When it is possible to find, between two neighboring cores, a good loge associated with a certain number n of electrons possessing a given organization of spin, it is said that there is a n-electron localized bond between these cores. In many cases it is not possible to do so. To find a good loge it becomes necessary to extend over more than two cores the space associated with the loge. When such a loge associated with n electrons is extended over p cores it is said that there is in the molecule a n-electron bond delocalized over these p centers.

Another situation which occurs is the existence of a loge associated with n electrons situated in the bond region but related with only one core. It is a lone n-electron loge.

A typical example of separation in loges is given in Figure 10. On atom A there is a lone pair; between A and B, a two-electron localized bond; and between B, C, and D, a four electron bond delocalized over three cores.

In principle, the separation of the space associated with a given molecule into good loges results from calculations based on the electronic wavefunction. Therefore to build the loges we need the wavefunction. But it is also possible to proceed in the opposite way. The analysis of experimental data, the comparison of a molecule with analogous simpler molecules, or the knowledge of a rough wavefunction, can give an idea of a good decomposition in loges. Starting from such a decomposition one can build a general theory to calculate electronic wavefunctions. For that purpose, only a topological description of the loges is needed.

The starting point of the general formulation was given by Daudel (232) in 1956.

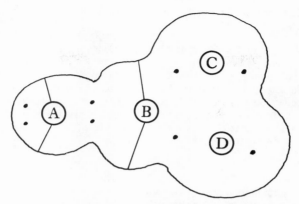

Fig. 10. Typical example of loge separation.

Let us take the example of Figure 10. If all cores are K shells it is natural to calculate approximate electronic wavefunctions starting from a generating function such as

$$\Upsilon = K_A(1, 2) \; K_B(3, 4) \; K_C(5, 6) \; K_D(7, 8) \; L_A(9, 10)$$
$$B_{AB}(11, 12) \; B_{BCD}(13, 14, 15, 16)$$

where the various functions K, L, and B are <u>loge functions</u>. These loge functions are also called group functions.

To transform the generating function into the wavefunction, we have to take account of the spin and of the Pauli principle. Let σ be a spin function, eigenfunction of S^2 and S_z; we shall write:

$$\Psi = \sum_p (-1)^p P \Upsilon \sigma$$

Following a variational method we shall calculate the loge functions in such a way that:

$$\delta < \Psi \mid H-E \mid \Psi > = 0$$

where H represents the exact total Hamiltonian of the molecule and E one of its energy levels.

To do the effective calculation it is necessary in practice to develop the various loge functions on the basis of monoelectronic functions χ as Slater orbitals or Gaussian functions. For example, an expansion like:

$$B_{BCD} = \sum_i \sum_j \sum_k \sum_l \chi_i(1)\chi_j(2)\chi_k(3)\chi_l(4)$$

will be used to represent the B_{BCD} loge function.

The concept of loge has been used to search for a theoretical justification of the existence of some molecular-property additivity rules. It is clear that such a rule for a certain property A will be understood if it is possible to associate with each loge i a component A_i in such a way that:

$$A = \sum_i A_i$$

and if a component associated with a certain kind of loge (for example a two-electron localized C–C bond) remains approximately constant within the limits of a given family of compounds.

Taking account of the definition of a loge, it is natural to define the fraction A_i of a molecular property associated with a loge i corresponding to a volume L_i with the help of the expression:

$$A_i = \int_{R^3 - L_i} d\tau_{p+1} \ldots d\tau_n \int_{L_i} \Psi^*(1, \ldots, n)\, \hat{A}\Psi(1, \ldots, n)\, d\tau_1, \ldots, d\tau_p$$

if there is a high probability of finding p electrons in the loge and if R^3 denotes the complete physical space.

It has been possible to show (232) that when the decomposition into loges is good the equation:

$$A = \sum_i A_i$$

is approximately satisfied if $\hat{A}$ is a monoelectronic operator. To recall the main steps of this demonstration let us consider a two-loge function such as:

$$\Psi = \mathcal{A} L_1(1, 2)\, L_2(3, 4, 5)$$

where $\mathcal{A}$ is an antisymmetrizer and where, to simplify the discussion, the spin is not considered explicitly.

If the loges are good, the functions L_1 and L_2 will be almost entirely localized in different parts of the space. Therefore an exchange integral such as

$$\int L_1^*(1,2) \, M_2(1,4,5) \, L_2^*(3,4,5) \, L_1(2,3) \, dv$$

will be negligible because

$$\int L_2^*(3,4,5) \, L_1(2,3) \, dv_3$$

is small. Then to calculate:

$$A = \sum_j \int \Psi^* \hat{M}_j \Psi \, dv$$

where $\hat{M}_j$ is a monoelectronic operator, we may use (with a good approximation) the simple product:

$$L_1(1,2) \, L_2(3,4,5)$$

which gives:

$$A = \int L_1^*(1,2) \sum_j {}_2^1 \, M_j L_1(1,2) \, dv$$

$$+ \int L_2^*(3,4,5) \sum_k {}_3^5 \, M_k L_2(3,4,5) \, dv$$

$$= A_1 + A_2$$

But this kind of relation can also be satisfied in some other cases such as the Faraday effect (237). The molecular rotativity $[\Omega]_M$ can be calculated effectively according to the following expression:

$$[\Omega]_M = \sum_{\text{bonds}} [\Omega]_{\text{bond}} + \sum_{\substack{\text{lone} \\ \text{pairs}}} [\Omega]_{\text{lone pair}} + \sum_{\substack{\text{atomic} \\ \text{cores}}} [\Omega]_{\substack{\text{atomic} \\ \text{core}}}$$

where the bonds may be either localized or delocalized.

There is another important point concerning the properties of loge functions. As they mainly describe local parts of atoms or molecules they can show the existence of various different local behaviors of the spin properties of the electronic density.

Let us take as an example the case of the beryllium atom. To write a more explicit form of the electronic wavefunction we can use Young operators (246).

If we consider a state corresponding to an eigenvalue of $2 h^2/4\pi^2$ for S^2 and $h/2\pi$ for S_z we must use for the space:

<table>
<tr><td>1</td><td>2</td></tr>
<tr><td>3</td><td></td></tr>
<tr><td>4</td><td></td></tr>
</table>

$\Phi_1 = \Lambda(1,2)\ \Lambda'(3,4) - \Lambda(2,3)\ \Lambda'(1,4) + \Lambda(2,4)\ \Lambda'(1,3)$

<table>
<tr><td>1</td><td>3</td></tr>
<tr><td>2</td><td></td></tr>
<tr><td>4</td><td></td></tr>
</table>

$\Phi_2 = \Lambda(1,3)\ \Lambda'(2,4) - \Lambda(2,3)\ \Lambda'(1,4) + \Lambda(3,4)\ \Lambda'(1,2)$

<table>
<tr><td>1</td><td>4</td></tr>
<tr><td>2</td><td></td></tr>
<tr><td>3</td><td></td></tr>
</table>

$\Phi_3 = \Lambda(1,4)\ \Lambda'(2,3) - \Lambda(2,4)\ \Lambda'(1,3) + \Lambda(3,4)\ \Lambda'(1,2)$

where Λ is symmetric and Λ' antisymmetric; and for the spin:

<table>
<tr><td>1</td><td>3</td><td>4</td></tr>
<tr><td>2</td><td></td><td></td></tr>
</table>

$\zeta_1 = \eta(1,2)\ \alpha(3)\ \alpha(4)$

<table>
<tr><td>1</td><td>2</td><td>4</td></tr>
<tr><td>3</td><td></td><td></td></tr>
</table>

$\zeta_2 = \eta(1,3)\ \alpha(2)\ \alpha(4)$

<table>
<tr><td>1</td><td>2</td><td>3</td></tr>
<tr><td>4</td><td></td><td></td></tr>
</table>

$\zeta_3 = \eta(1,4)\ \alpha(2)\ \alpha(3)$

where: $\eta(i,j) = \alpha(i)\ \beta(j) - \beta(i)\ \alpha(j)$

If we do the direct product of the space basis by the spin basis and extract the function corresponding to the totally anti-symmetric representation we obtain:

$$\Psi = \Phi_1\zeta_1 - \Phi_2\zeta_2 + \Phi_3\zeta_3$$

Now, let us consider the special case where the Λ's are completely localized loge functions, that is to say:

$$\exists L \in R^3,\ \forall M_i, M_j \notin L \Longrightarrow \Lambda(M_i, M_j) = 0$$

$$\exists L' \in R^3,\ L \cap L' = \emptyset,$$

$$\forall M_i, M_j \notin L' \Longrightarrow \Lambda(M_i, M_j) = 0$$

Therefore if: $M_1, M_2 \in L$ and $M_3, M_4 \in L'$, then $\Psi(M_1, M_2, M_3, M_4)$ reduces to:

$$\Psi(M_1, M_2, M_3, M_4) = \Lambda(M_1, M_2)\, \eta(\omega_1, \omega_2)\, \Lambda'(M_3, M_4)$$
$$\alpha(\omega_3)\, \alpha(\omega_4)$$

The atom is locally singlet in the loge L, and locally triplet in the loge L'. More precisely, the space function is symmetric in the loge L and antisymmetric in the loge L'. Furthermore as $\Lambda'(M, M) = 0$

$$\Psi(M_1, M_2, M, M) = \Lambda(M_1, M_2)\, \eta\, \Lambda'(M, M)\, \alpha\alpha = 0$$

and therefore, because of the antisymmetry of Ψ with respect to any permutation of the space and spin coordinates associated with two electrons:

$$\Psi(M_1, M, M_2, M) = -\Psi(M_1, M_2, M, M) = 0$$
$$= -\Psi(M, M_1, M_2, M)$$

The probability of finding two electrons in an elementary volume dv of the loge L' is zero.

Finally it is seen that the behavior of the electronic density must be rather different in the K loge and in the L loge. In the first one the probability of finding two electrons at a short distance can be great. This is not the case in the L loge, even if they have opposite spins. If the two loge functions are not completely localized, the difference of behavior will be reduced but obviously will stay significant so far as the loges are good loges.

2. Two-electron functions

The notion of loges is not the only way to introduce polyelectronic functions in a wavefunction. In fact two-electron functions have been introduced by Fock (238) in 1950, and by Hurley et al. (239) (see also Kapuy (240)) in 1953. The independent-pair model considered by Daudel and Durand (259) leads to a rather different kind of two-electron function. Let us consider, for example, a molecule with an uneven number 2p of electrons in a state where the z projection of the spin is zero. At a given instant we shall find p electrons with the z spin component $+\frac{1}{2}\dfrac{h}{2\pi}$ and p electrons with the z spin component $-\frac{1}{2}\dfrac{h}{2\pi}$. We know that two electrons with the same spin tend to stay far apart but that two electrons with opposite spin may be found in the same small region of space. A simplified picture of the molecule is obtained by assuming that it consists of a set of pairs of interacting electrons of opposite spin but that there is no coulombic interaction between the pairs.

The simplified Hamiltonian of the problem is the sum of p Hamiltonians h_{ij} representing the motion of a given pair ij in the field of the nuclei:

$$h_{ij} = h_i + h_j + e^2/r_{ij}$$

As in the independent-electron model, we are led to consider such equations as:

$$h_{ij} B_k(ij) = \epsilon_k B_k(ij)$$

and a product of p functions B is a rigorous solution of the corresponding Schrödinger equation.

If now we introduce the indistinguishability of the electrons, we are led to the wavefunction:

$$\Psi = \sum_p (-1)^p P \Upsilon \sigma$$

with

$$\Upsilon = B_k(1, 2) B_{k+1}(3, 4) \ldots B_{k+p-1}(2p-1, 2p)$$

Now as in the Hartree-Fock problem we can use the minimum number of B functions, or as in the extended HF scheme, the largest one. For certain states that minimum number is just one. Therefore the corresponding generating function is:

$$\Upsilon = B(1, 2) B(3, 4) \ldots B(2p-1, 2p)$$

The corresponding two-electron function B has been called biorbital by Bratoz (260) who has been the first to use such functions in atomic and molecular physics.

This kind of function has been used to study beryllium (261) and conjugated molecules (262). We shall not go into details because until now this method has not been used to describe the σ part of the molecules. But we must point out that, as the total space of the molecule is described by only one B function, it must be completely delocalized over that total space and therefore a biorbital is far from being a two-electron loge function. Furthermore we may add that the biorbitals have been shown to be able to take account of about 50% of the correlation energy.

In the opposite approach the maximum number of B functions is introduced. The name <u>geminal</u> is usually given to the corresponding two-electron functions. But there are many kinds of geminals and therefore many methods of calculating wavefunctions based on geminals. The method recently proposed by Ciampi and Paoloni (255) is perhaps the technique which derives most naturally from the independent-pair model. The basic assumptions are: (1) to consider only the n and the π electrons; (2) to evaluate explicitly only the interactions among paired electrons; (3) to write down the N-electron wavefunction as an antisymmetrical product of N/2 two-electron functions; (4) to express a geminal as a linear combination of products of Hückel-type MO's chosen in order to represent electrons in one doubly occupied orbital and in two singly occupied MO's (excited electron pairs).

Therefore the geminals must obey an equation of the type:

$$H(1,2)\ \Phi_R(1,2) = \epsilon_R \Phi_R(1,2)$$

In this equation:

$$H(1,2) = h(1) + h(2) + e^2/r_{12}$$

The total wavefunction is written as:

$$\Psi(1,2,\ldots,N) = \mathcal{A}_N[\Phi_1(1,2)\ \ldots\ \Phi_{N/2}(N-1,N)]$$

the geminals Φ containing a spin function.

To calculate the matrix elements of $H(1,2)$, various approximations are introduced. The matrix elements of $h(1)$ and $h(2)$ are identified with the classical α and β. Furthermore the matrix elements of e^2/r_{12} are calculated with proper two-electron integrals, for which the zero differential overlap (ZDO) approximation may be introduced.

Starting from this basis the authors have built a semiempirical approach and a parametric one. The first approach has been applied to the calculation of energy levels of N_2, C_2H_2, C_2H^-, and HCN. The parametrical approach has been used to discuss the possibility of existence of an hypothetical molecule: the diazabicyclo-octatriene or diazabarrelene:

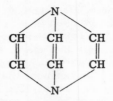

But there are much more general ways of using geminals. Let us forget the independent-pair model and also for a moment the concept of loge. We can consider the generating function:

$$\Upsilon = B_k \cdots B_{k,p-1}$$

as a simple trial function and try to find the best B's by following any kind of convenient variational procedure. We may or may not add some requirements for the geminals. If no additional requirement is introduced, the geminals are said to be free geminals (241, 242). But we can also ask them to satisfy a soft orthogonality condition such as

$$\int B_k(1,2) \, B_{k'}(1,2) \, dv_1 \, dv_2 = 0$$

or a strong orthogonality condition such as:

$$\int B_k(1,2) \, B_{k'}(1,3) \, dv_1 = 0$$

Orthogonal geminals possess interesting properties (244, 245). When the strong orthogonality condition is satisfied, it is often said that the separated-pair model is used and very often the geminals are more or less localized in various parts of the space of the atom or of the molecule considered.

But we must add another requirement in trying to find the most localized geminal possible, the notion of localizability being refined by using one of the classical criteria. In such a case the connection between the geminals and the loges becomes obvious: the geminals are simply loge functions.

Finally we can constrain the geminals to be by construction completely localized in completely different volumes, for example, into different loges. In that case, they represent a system in which there is a probability of 100% of finding two electrons in each loge. This model, which is extreme and does correspond to a physical situation, is however interesting to define some aspects of the correlation or as a starting point for more elaborate calculations.

Effective calculations have been made for the ground state of beryllium with free geminals, orthogonal geminals, and completely localized loge functions. Table 14 shows the values obtained. It is seen that all methods give satisfactory results. The quality of the wavefunction depends mainly on the number of monoelectronic functions introduced in the expansion of the two-electron loge functions. The quality of the wave function obtained by starting from completely localized loge functions is particularly significant; it shows that the correlation between the two loges is small.

Table 14. Ground state energy of beryllium (in atomic units)

Method	Energy	Reference
Orthogonalized geminals	$-14,611$	242
	$-14,6575$	247
Free geminal	$-14,621$	242
	$-14,65026$	248
Completely localized		
loge functions	$-14,5736$	243
Experimental value	$-14,667306$	

Various molecules have been studied by using geminals with strong orthogonality conditions. It is often said that in such cases the "separated-pair model" is used. Ahlrichs and Kutzelnigg (257) have applied the method to LiH, BeH_2, BH_3 and CH_4. In the case of LiH the results are highly accurate. In the other cases the results are not so spectacular but the method does appear to offer a significant improvement over the independent-particle model, at little extra cost in time and effort, both as an end in itself, and as a starting point for more accurate calculations. Scarzafara (258) has used the same model to determine the electronic wavefunction associated with the ground state of the water molecule as a starting point. Then a configuration-interaction calculation is made, employing the natural orbitals of the separated-pair wavefunction as a basis, with the separated-pair wavefunctions as the leading configuration. The corresponding correlation energy is partitioned into various intrageminal and intergeminal components. From this analysis it is concluded that for the water molecule the separated-pair model is capable of yielding approximately 50% of the correlation energy.

Actually this result is understandable and was predicted some time ago by the Sinanoğlu theory of electron correlation in atoms and molecules (c.f. for example reference (187) and earlier references cited therein). In that theory, the N-electron system for ground states is decomposed into $N(N-1)/2$ pair correlations. These pair correlations are calculable by individual variational principles. Such calculations have been carried out in the past by Sinanoğlu and co-workers in atoms and by other workers more recently. In that theory inter- and intraorbital correlation occur on the same footing, as the mathematical derivations rigorously show. Also in the Sinanoğlu theory the pairs of electrons refer to different pairs of either molecular orbitals or to localized orbitals. The energy and the wavefunction are transformed from one description to the other quite rigorously by unitary transformations.

Silver (252) has suggested another way to use two-electron functions in studying molecules. His method is based on a natural orbital expansion of interacting geminals. Effective calculations have been done in the case of lithium hydride, boron hydride (253), and imidogen (254).

Another interesting approach is the procedure suggested by Klessinger and McWeeny (236). They introduced what they call "group functions." The group functions are a variety of loge functions. A minimal basis of Slater-type orbitals is used, the values of the exponents being:

$$\text{carbon} \quad \zeta \ (1\underline{s}) = 5.7 \quad \zeta \ (2\underline{s}) = \zeta \ (2\underline{p}) = 1.625$$
$$\text{hydrogen} \ \zeta \ (1\underline{s}) = 1.0$$

From this basis, a set of orthonormal orbitals is built: $\bar{h}_1$, $\bar{h}_2$, $\bar{h}_3$, $\bar{h}_4$ for the hydrogen atoms, $\bar{k}$ for the carbon K shell, and $\bar{t}_1$, $\bar{t}_2$, $\bar{t}_3$, and $\bar{t}_4$ for the tetrahedral hybrids. Then, the two-electron functions are written including a standard one-determinental function $| \ \bar{k}\alpha \ \bar{k}\beta \ |$ to represent the inner shell and four C–H bond functions. These bond functions are written as:

$$\Phi_1 = \sum_j c_{1j} \ \phi_{1j} \quad j = 1, 2, 3$$

where

$$\phi_{1_1} = (1/\sqrt{2}) \ | \ \bar{t}_1 \alpha \bar{t}_1 \beta \ |$$

$$\phi_{1_2} = (\tfrac{1}{2})\{|\bar{t}_1\alpha\bar{t}_1\beta| - |\bar{t}_1\beta\bar{t}_1\alpha|\}$$

$$\phi_{1_3} = (1/\sqrt{2})|\bar{h}_1\alpha\ \bar{h}_1\beta|$$

When self-consistency is achieved, the P matrix appears to be the following:

$\bar{h}_1$	$\bar{h}_2$	$\bar{h}_3$	$\bar{h}_4$	$\bar{k}$	$\bar{t}_1$	$\bar{t}_2$	$\bar{t}_3$	$\bar{t}_4$
0.92	0	0	0	0	0.98	0	0	0
	0.92	0	0	0	0	0.98	0	0
		0.92	0	0	0	0	0.98	0
			0.92	0	0	0	0	0.98
				2	0	0	0	0
					1.08	0	0	0
						1.08	0	0
							1.08	0
								1.08

The corresponding ground-state energy is −53.48 ev. There is a significant improvement in comparison with the standard SCF calculation made with the same basis (−53.44 ev). The dipole moment associated with each C−H bond is −1.86 D.

The kind of bond function which has been used to describe methane quantitatively is useful in discussing various problems even qualitatively. The clear distinction between covalent and dative bonds remains difficult. The main difficulty arises from the fact that it is usually said that a dative bond appears when during the formation of the bond one atom gives a lone pair to another atom; the covalent bond appearing when each atom contributes one electron to the bond. Such definitions are based on the mechanisms of bond formation. In some cases several mechanisms may be assumed for the same bond which leads to different conclusions concerning the nature of the bond. It is certainly better, in determining the nature of a bond, to search for actual structural differences than to take account of an hypothetical history of the formation of the bond. Recently (249) the notion of loge has been used to propose a distinction between covalent and dative bonds. Let us compare borazane $H_3B \leftarrow NH_3$, which contains a typical dative bond $B \leftarrow N$, and the hypothetical molecule H_2B-NH_2 where the B−N bond is considered to be covalent. (As this molecule is taken as a model of a covalent bond, the possible delocalization of the lone pair is neglected.) The B–N bond loge in borazane is extended between a first group of loges containing the B core and three two-electron loges B–H, and a second group of loges containing the N core and three two-electron loges N–H. When there is in each loge the most probable number of electrons (two), the total charge of the first group of loges is zero; the second group possesses the total charge + 2. We can say that dative bond B $\leftarrow$ N is established between a group of loges possessing the <u>most probable charge</u> zero and another possessing the most probable charge + 2. On the other hand, it is easy to see that in H_2B–NH_2, the B–N bond loge (covalent bond) is established between two groups of loges possessing the same most probable charge + 1.

If we now use the kind of two-electron loge functions introduced for methane, the B–N loge function will be:

$$\Phi_{BN} = a \mid \bar{t}_B \alpha \; \bar{t}_B \beta \mid + b \left\{ \mid \bar{t}_B \alpha \; \bar{t}_N \beta \mid - \mid \bar{t}_B \beta \; \bar{t}_N \alpha \mid \right\}$$
$$+ c \mid \bar{t}_N \alpha \; \bar{t}_N \beta \mid$$

where $\bar{t}_B$ and $\bar{t}_N$ denote convenient hybrid orbitals associated with boron and nitrogen.

In the case of H_2B–NH_2, as the B–N loge lies between two groups of loges possessing the same most probable charge the difference between a and c in Φ_{BN} will be rather small because it is only due to the difference of the electronegativities of boron and nitrogen. But in the case of H_3B $\leftarrow$ NH_3, as the NH_3 group of loges has a most probable charge of +2 and the H_3B group a corresponding charge of zero, the coefficient c will be much greater than the coefficient a. Therefore we must anticipate that the gravity center of the electronic charge associated with the B–N dative bond will be much nearer the nitrogen atom than it is in the B–N covalent bond.

Similar results have been found quantitatively following another method in the case of some N–O bonds (250). Therefore a dative bond can be considered from two points of view. If for example we consider the formation of the B–N bond in borazane starting from the two groups BH_3 and NH_3, the formation of the bond corresponds to an electron transfer from the nitrogen to the boron. This transfer has been clearly shown recently by Veillard and Daudel (256), studying the differential electronic density. This is the phenomenon which is represented by the classical notation H_3N^+ $\rightarrow$ $^-BH_3$. However in the B–N loge, the gravity center of the electronic charge lies nearer the nitrogen, that is to say, in the opposite direction with respect to the polarity indicated by this notation.

3. The LCBO method

We have seen that, for a molecule in which the number of electrons commonly found outside of the core regions is twice the number of adjacent atomic cores, a good division in loges consists of regions extended between two such adjacent cores in which there is a high probability of finding two electrons with opposite spins. It is easy to see that this is so in the case of paraffins. Therefore a convenient generating function for such a molecule is:

$$\Upsilon = K_1(1,2) \ K_2(3,4) \ . \ . \ . \ . \ L_{C_1H_a}(p, p+1)$$

$$L_{C_1H_b}(p+2, p+3) \ . \ . \ . \ . \ L_{C_1C_2}(q, q+1)$$

$$L_{C_2C_3}(q+2, q+3) \ . \ . \ . \ .$$

where the K are the various carbon core functions, the L_{CH} are the loge functions associated with the C$-$H bonds and the L_{CC} are the loge functions associated with the C$-$C bonds. Following the method described in section 1 of this chapter, the various two-electron functions have to be expanded on a basis of monoelectronic functions:

$$L = \sum_i \sum_j \chi_i \chi_j$$

Now if that expansion is limited to the first term and if the core is not included in the wave function, the linear combination of bond orbitals (LCBO) method is obtained. The generating function Υ is now:

$$\Upsilon = 1s_1(1) \ 1s_1(2) \ 1s_2(3) \ 1s_2(4) \ . \ . \ .$$

$$\chi_{C_1H_a}(p) \ \chi_{C_1H_a}(p+1) \ \chi_{C_1H_b}(p+2) \ \chi_{C_1H_b}(p+3) \ . \ . \ .$$

$$\chi_{C_1C_2}(q) \ \chi_{C_1C_2}(q+1) \ \chi_{C_2C_3}(q+2) \ \chi_{C_2C_3}(q+3) \ . \ . \ .$$

If we are concerned with the ground state of a paraffin (which is a singlet state), the corresponding wavefunction:

$$\Psi = \sum_P (-1)^P \, P \, \Upsilon \sigma$$

is readily found to be:

$$\Psi = \det 1s_1(1) \, 1s_1(2) \, \overline{1s_2}(3) \, \overline{1s_2}(4)$$

$$\ldots \chi_{C_2 C_3}(q+2) \, \overline{\chi}_{C_2 C_3}(q+3) \ldots$$

The various χ's are called bond orbitals. Let T be a unitary transform. We can replace the bond orbitals in the determinant by the following linear combinations of them:

$$\phi_j = \sum_i c_{ij} \, \chi_j$$

The c_{ij} correspond to the unitary transform T. The determinant (and therefore the wavefunction Ψ) will remain the same. We can write:

$$\Psi = \det \phi_1(1) \, \bar{\phi}_1(2) \, \phi_2(3) \, \bar{\phi}_2(4) \ldots \bar{\phi}_n(2n)$$

If now the condition:

$$\delta < \Psi \mid H - E \mid \Psi > = 0$$

is taken into account, it is readily seen that the coefficients c_{ij} are solutions of Roothaan's equations.

Hall (251) proposed a semiempirical method based on that formalism. Let us assume that the self-consistent field operator h^{SCF} is known and <u>fixed</u>. It is possible to calculate the c_{ij}'s with the help of the McDonald theorem. Then, the best c_{ij}'s are solutions of the secular system

$$\sum_j c_{ij} [< \chi_i h^{SCF} \chi_j > - \epsilon < \chi_i \chi_j >] = 0 \quad (i = 1 \ldots n)$$

Since we saw that the paraffin bonds are well individualized, they must correspond to good loges and therefore, to a first approximation, the overlap integrals can be neglected:

$$< \chi_i \chi_j > = 0 \quad (i \neq j)$$

Let us consider, as an example, the methane molecule. In that case symmetry considerations lead to the following remarks. The expression $< \chi_i h \chi_j >$ (with $i \neq j$) does not depend on the indices. This is also the case for $< \chi_i h \chi_i >$. Hence we can set:

$$\alpha_{CH} = <\chi_i h \chi_i>$$
$$\beta_{CH, CH} = <\chi_i h \chi_j> \qquad i \neq j$$

Putting:

$$\alpha_{CH} = \alpha \qquad \beta_{CH, CH} = \beta$$

the secular equation is written as:

$$\begin{vmatrix} \alpha-\epsilon & \beta & \beta & \beta \\ \beta & \alpha-\epsilon & \beta & \beta \\ \beta & \beta & \alpha-\epsilon & \beta \\ \beta & \beta & \beta & \alpha-\epsilon \end{vmatrix} = 0$$

The roots of this equation are:

$$\epsilon_1 = \alpha + 3\beta$$

and

$$\epsilon_2 = \alpha - \beta \text{ (triple root)}$$

Following Koopmans' theorem, these roots give approximations to the ionization energies. If we want to reproduce the experimental values (13ev and 20ev) we are led to the following values for the parameters:

$$\alpha = -14.75ev \qquad \beta = -1.75ev$$

which can be used to evaluate ionization energies in other paraffins. But to do such calculations the values of other parameters are needed:

$$c = <\chi_{CC'} h \chi_{CC'}>$$
$$d = <\chi_{CC'} h \chi_{CH}>$$
$$e = <\chi_{CC'} h \chi_{C'C''}>$$

Finally Hall has determined the values of the five parameters α, β, c, d, and e which give the best fit over eight molecules. Table 15 compares experimental and theoretical data.

Brown (263) has used the same kind of method to determine the atomization energies of saturated hydrocarbons. Overlap integrals are included in the calculations. Brown sets:

$$S = <\chi_{CH} \chi_{CH'}> \qquad \alpha = <\chi_{CH} \chi_{CH}>$$
$$\chi + h\gamma = <\chi_{CC'} \chi_{CC'}> \qquad \beta = <\chi_{CH} h \chi_{CH'}>$$

$$\gamma = \beta - S\alpha \qquad \theta\beta = \langle \chi_{CH} \chi_{CC'} \rangle$$
$$\eta\beta = \langle \chi_{CC'} \, h \, \chi_{C'C''} \rangle \qquad \eta S = \langle \chi_{CC'} \chi_{C'C''} \rangle$$

Table 15. First ionization energies of paraffins (in ev)

	Calculated	Measured
Propane	11.214	11.21
Pentane	10.795	10.80
Heptane	10.323	10.35
Nonane	10.224	10.21
Decane	10.194	10.19

Thus, a proportionality between the overlap integral and the resonance integrals is assumed. Furthermore the total energy is obtained by taking the sum of the orbital energies. We know that this is a rough approximation. Table 16 shows some results which are obtained by solving the corresponding secular equations. It turns out that if S is neglected the atomization energy can be written:

$$N_{CH}(2\alpha) + N_{CC}(2\alpha + 2h\gamma)$$

If N_{CH} and N_{CC} represent respectively the number of C–H and C–C bonds.

Table 16. Expressions of atomization energies

Methane	$8\alpha - 24\gamma S$
Propane	$20\alpha + 4h\gamma - (28 + 40\theta^2 + 4\eta^2)\gamma S$
Butane	$26\alpha + 6h\gamma - (32 + 56\theta^2 + 8\eta^2)\gamma S$
Isobutane	$26\alpha + 6h\gamma - (36 + 48\theta^2 + 12\eta^2)\gamma S$

If the overlap integral S is taken into account it is possible to evaluate isomerization energies. For example this energy appears to be:

$$\Delta E = (4 - 8\theta^2 + 4\eta^2)\gamma S$$

when the following pairs of isomers are compared: butane, isobutane; pentane, isopentane; hexane, isohexane. The experimental values of ΔE are 1.7, 1.9, and 1.7 kcal/mole respectively. The fact that ΔE is about the same for all pairs is therefore in agreement with the theoretical expression. Furthermore if the two following pairs are considered: normal pentane and 2,2-dimethyl propane; hexane and 2,2-dimethylbutane, the theory leads to the following expression for the isomerization energy ΔE:

$$\Delta E' = (12 - 24\theta^2 + 12\eta^2)\gamma S$$

Therefore:

$$\Delta E' = 3 \Delta E$$

Again, this result agrees with experiment, as $\Delta E'$ is found to be 4.7 kcal/mole. Brown used the same procedure to evaluate with success various dissociation energies.

To establish a bridge between the LCBO method and the use of two-electron functions, let us show that it is possible to treat the same problems by using two-electron loge functions (264).

The generating function will remain:

$$\Upsilon = K_1(1, 2) \, K_2(3, 4) \ldots \ldots L_{C_1 H_a}(p, p+1) \ldots \ldots$$
$$L_{C_1 C_2}(q, q+1) \ldots \ldots$$
$$= \prod_r \mathscr{L}_r$$

if the general notation $\mathscr{L}_r$ is given to any two-electron loge function. Following the idea of Ludeña and Amzel (243), we shall use completely localized loge functions. Let V_r be the volume in which the function $\mathscr{L}_r$ is localized. We can write:

$$\exists \{V_1, \ldots, V_r, \ldots V_n\} \in R^3, \; V_r \cap V_{r'} = \phi,$$
$$\forall M \notin V_r \Longrightarrow \mathscr{L}_r(\ldots M \ldots) = 0$$

Therefore $\Psi = \sum_P (-1)^P P \Upsilon \sigma$ reduces to:

$$\Psi = \prod_r \mathscr{L}_r^\tau$$

where $\mathscr{L}_r^\tau$ represents localized antisymmetrized loge functions. Let us set as usual:

$$H = \sum_i h_i + \frac{1}{2} \sum_{i,j} e^2/r_{ij}$$

and:

$$\bar{\alpha}_r = \sum_i \langle \mathcal{L}_r^\tau | h_i | \mathcal{L}_r^\tau \rangle$$

$$\bar{\gamma}_{rr} = \frac{1}{2} \sum_{ij} \langle \mathcal{L}_r^\tau | e^2/r_{ij} | \mathcal{L}_r^\tau \rangle$$

$$\bar{\gamma}_{rr'} = \frac{1}{2} \sum_{ij} \langle \mathcal{L}_r^\tau | e^2/r_{ij} | \mathcal{L}_{r'}^\tau \rangle \text{ for } r \neq r'$$

The total energy is readily expressed as:

$$E = \sum_r [\bar{\alpha}_r + \bar{\gamma}_{rr}] + \sum_{r < r'} \bar{\gamma}_{rr'}$$

To evaluate the isomerization energy, ΔE, it is sufficient to consider the two parts of the paraffins which differ, that is to say,

normal molecule   iso-derivative

Therefore no calculations are necessary, it is only necessary to look at the structure of the molecule to find:

$$\Delta E = [3\bar{\gamma}_{CC,CC} + 3\bar{\gamma}_{CH,CH} + 6\bar{\gamma}_{CC,CH}]$$

$$- [2\bar{\gamma}_{CC,CC} + 2\bar{\gamma}_{CH,CH} + 8\bar{\gamma}_{CC,CH}]$$

$$= [\bar{\gamma}_{CC,CC} + \bar{\gamma}_{CH,CH} - 2\bar{\gamma}_{CC,CH}]$$

if only the interactions between two adjacent loges are introduced as in the usual LCBO calculations.

In the same way, it is readily seen that:

$$\Delta E' = 3 [\bar{\gamma}_{CC,CC} + \bar{\gamma}_{CH,CH} - 2\bar{\gamma}_{CC,CH}]$$

$$= 3 \Delta E$$

It is interesting to point out that using this more elaborate theory, which includes intra-loge electronic correlation, the derivation of the preceding formulae is much simpler and needs no calculation. Furthermore its pure topological background is more obvious. We believe that the completely localized loge functions method could be an excellent starting point for more elaborate calculations. It would certainly be very interesting to inject this method into the powerful algorithm recently proposed by Diner, Malrieu, and Claverie (265).

The energies of different molecules can also be broken down into different contributions some of which can be evaluated semiempirically according to the Sinanoğlu theory. ((187) and earlier references.) For example, the binding energy of a molecule is quite naturally given as the sum of a Hartree-Fock binding and a correlation binding part. With a ZDO approximation made on the correlation part, the correlation binding part separates into intra- and interatomic correlations more rigorously. Without the ZDO approximations the correlation energy can also be written as the bond and interbond parts. This theory though proposed some time ago (for example, see Sinanoğlu in Modern Quantum Chemistry Vol. II, Academic Press, 1965) has been implemented more recently for quite a number of molecules by Brown and Roby (289), Pamuk and Sinanoğlu (196).

Lennard Jones and Hall (266) have studied the distribution of electric charge in positive paraffin ions using essentially the method of linear combination of bond orbitals. Analogous to what is done in the LCAO method, the quantity c_{ij}^2 in the expansion:

$$\phi_j = \sum_i c_{ij} \chi_i$$

can be regarded as the contribution of orbital ϕ_j to the electron population of bond i, provided the overlap integrals are neglected.

Now, considering a positive ion of a paraffin resulting from this by the "loss of an electron" from orbital ϕ_j, the ionization produces a hole in the electron population in bond i equal to c_{ij}^2. Such a hole corresponds to a positive charge of

$$ec_{ij}^2$$

By solving the secular equation with the parameters already obtained, Lennard Jones and Hall found a positive charge of:

0.035 in bond 1	0.200 in bond 3
0.115 in bond 2	0.234 in bond 4

in the case of normal octane.

The effect of ionization is much more marked in the center of the molecule than at its ends. Thompson (267), Coggeshall (268), and Lorquet (269) have used this kind of result to explain the nature of positive ions which are produced when paraffins are irradiated with electron beams.

The LCBO method has been used by Julg (270) to analyze the effect of a substituent (like a fluorine atom) on the distribution of electronic charge in a saturated hydrocarbon. Julg showed that the populations of the C—C bonds are either all increased or all decreased depending on the electronegativity of the substituent. Hence the effect is not oscillatory. These results generalize those of Sandorfy and Daudel described in Chapter 1.

Chapter 4

Recent Developments

Well over a hundred papers were published during 1969 on σ-electron problems. Among these a large number were applications of known methods to individual molecules or phenomena. We shall mention some of these briefly. More attention will be paid to a few recent publications which are of general importance.

1. Hückel methods

In 1968 Harris (271) made an interesting comment on self-consistent Hückel methods, that is the iterative Hückel, or "ω" technique. He called attention to the fact that a self-consistent Hückel theory should be characterized by a set of equations different from those which appear to have been employed by some investigators. "The point which has been overlooked is that the charge dependence of the parameters must be included in the differentiations which define the minimum-energy condition and thereby the molecular orbitals. This means that it is insufficient to solve conventional Hückel-theory equations iteratively until the solution is consistent with the parameters, for such a solution will not necessarily correspond to a charge distribution of minimum energy."

The total energy for a closed-shell state in Hückel theory is written as

$$E = 2\sum_{i=1}^{n} E_i$$

$$= 2\sum_{\mu,\nu=1}^{N} \sum_{i=1}^{n} C_{\mu i}^{*} \ C_{\nu i} \ H_{\mu\nu}$$

$$= \sum_{\mu,\nu=1}^{N} P_{\mu\nu} \ H_{\mu\nu}$$

91

where E_i is the energy of an orbital, the $C_{\mu\nu}$ are the usual LCAO MO coefficients and

$$P_{\mu\nu} = 2 \sum_{i=1}^{n} C_{\mu i}^{*} C_{\nu i}$$

In a self-consistent calculation, the $H_{\mu\nu}$ depend upon the α and β as well as on the C_μ ; by minimizing E subject to the condition that the molecular orbitals are mutually orthogonal, and by using the method of Lagrangian multipliers, we are led to the expression

$$F_{\mu\nu} = H_{\mu\nu} + \sum_{\mu,\nu=1}^{N} P_{\mu\nu} \frac{\partial H_{\mu\nu}}{\partial P_{\mu\nu}}$$

where $F_{\mu\nu}$ is an element of the matrix of the SCF operator F which defines the molecular orbitals. The relation between the two operators is given by

$$H = F - G/2$$

where G is the electronic interaction operator. F contains the electronic interaction energy twice.

In the ω technique the $H_{\mu\mu}$ are supposed to depend on the $P_{\mu\mu}$. Harris evaluated the $F_{\mu\nu}'$ for given $H_{\mu\nu}$ and showed the inadequacy of the usual iterated Hückel equations. Jug (272) studied the relationship between F and H through an expansion technique and examined the Wolfsberg-Helmholtz (WH) approximation. He found this approximation improper if applied to the $F_{\mu\nu}$ directly but proper if it is incorporated into F via introduction into H.

In a later paper Jug (273) showed that consistency requirements for the α parameters led to an explanation of the WH approximation for β, and discussed the dependence of α and β on distance in diatomic molecules.

Corrington and Cusachs (274) also examined the usual procedures for evaluating charge separation in heteronuclear molecules. They point out that "iteration to charge self-consistency constitutes an improvement only when proper account is taken of the potentials due to charges on neighbor atoms, which may be of the same order of magnitude as the effects of one-center charge loss or gain."

Miller and Cusachs (275) examined the applicability of semi-empirical MO methods to bonding in sulfur compounds starting with S_6 and S_8. They found that $\underline{d}$ orbitals are unimportant in the ground state but are important in excited states. They used overlap-matched Slater-type atomic orbitals and Cusachs' formula for the off-diagonal matrix elements. Their orbital exponent for S $3\underline{d}$ was 1.7.

Ferreira and Bates (276) examined the correspondence between the ω technique and "differential ionization energies methods" (see Klixbüll-Jörgensen et al. (74)) and showed that they are equivalent if correlation energy is neglected.

Among recent works relating to special problems we mention a paper by Kaufman and Harkins (277) on the reaction of hot carbon atoms with O_2 which they studied with the EH method. Hopkinson et al. (33) applied the same method and a nonempirical SCF method to the protonation reactions of acids, amides, and esters. Chesnut and Mosely (278) studied the geometries of molecular complexes by an EH approach. Yeranos (279) computed bond dissociation energies by a slightly modified EH approach and de Jeu and Benader (280) compared the Hoffmann and Pople-Santry approximations in relation to nuclear spin-spin coupling constants. Kettle and Tomlinson (281) studied the electronic structure of boron hydrides, Kier and George (282) the conformations of aminoacides, Rossi, David, and Schor (283) polypeptide chains, McAloon and Webster (284) a dimer model for the hydrated and the ammoniated electron, Olsen and Burnelle (285) the structure of the radical CO_3; all these investigators used the EH method. Boyd (286) used a simulated (Lipscomb-type) method for studying the components of adenozine triphosphate.

2. Pariser-Parr-Pople methods

We have dealt at some length with ideas of Cook, Hollis and Mc-Weeny (section 7 of chapter 2) who showed how the results of an ab initio calculation can be reproduced with fair accuracy if a "natural" set of symmetrically orthogonal hybrids is used as a basis. The appropriate choice of such a basis, adapted to the nature of the problem, makes certain integrals which are usually neglected in the ZDO approximation truly small while other integrals can be scaled according to an empirical procedure.

Cook and Palmieri (287), who continued this analysis, made an attempt to eliminate the shortcomings of this scaling procedure. In fact, it is not always easy to find a "natural" basis set and the scaling depends on the basis. This is particularly true if the variation of internuclear distances is desired. Thus Cook and Palmieri investigated approximation methods in order to obtain the modified ZDO integrals over the orthogonal basis set without empirical scaling so that the method becomes "ab initio" but still requires relatively simple computer calculations. They proposed a method which consists of computing the one-electron Hamiltonian in the orthogonal basis and the two-electron integrals in a basis where the Slater-type orbitals are expressed through a minimal linear Gaussian expansion. This "mixed basis" method gave excellent agreement with the full STO calculation for all cases they investigated (σ, π, and three-center bonds) which included molecules composed from both first and second row elements. An extension to nonminimal basis sets is possible.

In a paper on "the performance and parameter problems of approximate molecular orbital theory . . . ," Roby and Sinanoğlu (288) considered the problems that arise after a theoretically correct breakdown of the molecular wave function into Hartree-Fock and electron-correlation parts. They examined mainly the NDDO method of Pople et al. (98) and the "many-center zero differential overlap" (MCZDO) method of Brown and Roby (289), but they also considered some more approximate ZDO methods. They divided the problems into four categories: (1) The "intramolecular environment problem" comes from the need of taking into account the effects of the intramolecular environment on the basis atomic orbitals which have been optimized for the isolated atoms. There they considered a version of Brown and Heffernan's (84) VESCF technique. In this technique (discussed in section 1 of chapter 2), one-center integrals are made dependent on the effective nuclear charge, which in turn depends on the electron density through Slater's rules (for example). Two-center integrals were shown to depend only slightly on the effective nuclear charge. This technique can be employed only when single-exponent orbitals (STO) are used. (2) The "Hartree-Fock problem." In order to simulate as closely as possible the results of accurate HF calculations while preserving the advantages of single exponent Slater orbitals, Roby and Sinanoğlu suggest using the rules given by Burns (290) or Clementi and Raimondi (45) for calculating orbital exponents rather than Slater's rules, and using an HF average scaling scheme based on HF atomic orbitals and STO's for $\underline{s}$ orbitals. (3) The problem of transforming the integrals to a basis of Löwdin orthogonalized atomic orbitals can be solved empirically in Cook et al.'s manner (178) where only repulsion integrals are scaled, the one-center ones being increased by 12%, the two-center ones decreased by 12%. This factor can actually be varied to fit some chosen experimental property. (4) The electron correlation problem. The "shrunk core" and "pair populations" methods of Hollister and Sinanoğlu (291) and schemes by Brown and Roby (289) and Pamük and Sinanoğlu (196) were considered. The latter lead to the simple formulas

$$E_{Corr} \approx \tfrac{1}{4} \sum_{\mu\nu} P_{\mu\mu} P_{\nu\nu} \; \epsilon_{\mu\nu}^{Corr}$$

where the $P_{\mu\mu}$ are diagonal elements of the bond order matrix, and the $\epsilon_{\mu\nu}^{Corr}$ are pair correlation energies over atomic orbitals. (See equations 12 and 13 in (288).)

The CO molecule was used as a test case in the work of Roby and Sinanoğlu. They found the performance of the NDDO method encouraging but concluded that care is needed in interpretations "derived from calculations carried out at a more approximate level."

Westhaus and Sinanoğlu (292) in a recent paper examined the structure and transformation properties of correlation functions for open-shell states of molecules in nonorthogonal AO as well as in MO basis sets. This study may turn out to be a useful guide for semiempirical calculations.

A recent work by Jug (293) analyzes the CNDO method and its modifications. He has shown that the invariance requirements for atomic integrals under local rotations and local hybridization are not essential, but that the SCF equations have to be invariant under unitary transformations of the atomic basis set. This leads to the possibility of introducing different scale factors for the various $2\underline{s}$ and $2\underline{p}$ orbitals at the same center (cf. sec. 7, chap. 2). This is important for calculations of electronic spectra when both σ and π electrons are present. As already observed by Del Bene and Jaffe (210, 213) giving the same scale factor to σ and π orbitals leads to a complete mixing of σ and π levels. ("With too little shielding, the π-electron levels emerge too deeply into the range of the σ-electron levels.")

Jug emphasizes that Pople's invariance requirements do not imply the invariance of the integrals. The use of different scale factors makes it possible to obtain correct values for physical quantities in many cases where the original CNDO methods do not do this. This is even more true for $\underline{d}$ orbitals, as already pointed out by Santry and Segal (105).

A very interesting reexamination of the σ-π problem has been made by S. Fischer (294).

All the previous critical reviews of the foundations of approximate MO methods stressed the importance of the choice of the atomic orbitals basis. A suggestion of Rastelli and Del Re (295) may become important in this respect. These authors introduced single exponential functions of the forms

$$1\underline{\bar{s}} = 1\underline{s}, \quad 2\underline{\bar{s}} = N_1 \ (2\underline{s} + b \ 1\underline{s} + a \ 3\underline{s})$$
$$2\underline{\bar{p}} = N_2 \ (2\underline{p} + c \ 3\underline{p})$$

as an alternative to the Slater orbitals $1\underline{s}$, $2\underline{s}$, $2\underline{p}$ for the atoms of the first period. The constants a and c should be chosen in order to make the basis consistent with a given preliminary description of the molecule (296). These "promoted Slater orbitals" are intended for molecular calculations.

We should like to mention here an interesting work by Hegyi, Mezei, and Szondy (297) on "determination of molecular properties by the method of moments" which would be an alternative to energy variation and would lead to better results in certain cases.

Of great importance to all iterative methods are the studies of Paldus and Čížek (298) on the stability conditions for the solution of the HF equations.

The method of pseudopotentials which can be used instead of orthogonalization against the inner-shell orbitals seems to hold promise of applicability to larger molecules. Outside of the comprehensive treatise of Gombás (299) which contains many references to his earlier work and some of his more recent publications (300, 301), we mention in this respect the works of Gombás and Szondy on all the atoms (302) and on the Li_2 molecule (303); and those by Kutzelnigg (304) and McGinn (305, 306).

3. Parametrization of CNDO methods

Several authors made detailed studies to optimize the parametrization of the CNDO method or proposed modified versions of it in order to obtain good values for given physical properties.

Wiberg (307) applied the CNDO method to hydrocarbons, their cations, free radicals, and carbonium ions, and made a thorough investigation of the effect of the variation of parameters on the computed geometries.

Lo and Whitehead (308) developed a scheme to calculate heats of atomization, bond lengths, and resonance energies of benzenoid hydrocarbons taking the σ electrons into account. Other works of the same authors concerned acidic hydrocarbons (309), fulvenes (310), and related problems. Boyd and Whitehead (311) have shown that bonding parameters based on the dissociation energies of the first-row homonuclear diatomic molecules give more accurate bonding energies than the bonding parameters based on the dissociation energy of H_2. They applied their modified CNDO method with success to a large number of molecules.

Fischer and Kollmar (312) proposed another reparametrization of the CNDO method and obtained heats of atomization, bond lengths, bond angles, and force constants in good agreement with experiment.

An interesting idea came from Linderberg (313) who pointed out that the equivalence between dipole-length and dipole-velocity forms of oscillator strengths places a condition on the parameters used in the PPP method and leads to a new approach for the determination of β. Jug (273) has shown that it leads to an explanation of the WH form of β.

Davies (314) described a perturbation method for use with the CNDO/2 theory with applications to electric polarizabilities.

Jug (315) published an extensive review on "the development of semiempirical methods in the MO formalism" which contains a chapter on σ electrons.

4. Calculations on individual molecules

A partial list of papers containing calculations on individual molecules follows.

Dewar and Worley (316, 317) applied the MINDO (modified INDO) method to the interpretation of photoelectron ionization potentials for a large number of organic compounds; Baird and Dewar (318, 319) to the heats of formation of hydrocarbons. The parameters were chosen to fit observed heats of formation of reference molecules. Baird, Dewar, and Sustmann (320) extended this work to nitrogen and oxygen containing molecules.

Betsuyaku (321) studied the nitrite ion in the CNDO framework.

Segal et al. (322) computed vibrational intensities (dipole moment derivatives) by the CNDO/2 method in the case of F_2CO.

King et al. (323) computed the four lowest doublet states of the fluorosulfate radical.

Clark (324, 325) applied the CNDO/2 method to cyclopropane, ethylene oxide, and ethylenimine.

Yonezawa, Katô, and Katô (326) used a modified semiempirical procedure to treat the electronic structure and spin-orbit coupling in azabenzenes.

Becker and Dahl (327) worked with $TiCl_4$, Carbo et al. (328) with the borazarobenzenes by CNDO methods.

Basilevsky and Chlenov (329) computed a potential surface for the $CH_3^+ + CH_2 = CH_2$ reaction.

Pullman and Berthod (330) studied the tautomerism in isomeric oxypurines by both the CNDO/2 and the SCF MOCI method.

Cheesman et al.(331) used the VESCF MO approximation in the study of halogen-halogen bonding.

Tinland (332) gave CNDO results on acrolein.

The geometry of the CH_5^+ ion was investigated by Gamba et al. (333), Gole (334), and Ehrenson (335).

Works relating to photoelectron spectra come from Branton et al. (336) on NH_3 and ND_3, and from Asbrink et al. (337) on benzene.

Canadine and Hillier (338) examined the effect of ligand field terms in approximate MO calculations on metal complexes.

We close this list with the very informative comparative study of A. Pullman (339) of different all-valence electron calculations on biological purines and pyrimidines.

5. Calculations on electronic spectra

Jungen and Labhart (340) used an interesting PPP π-σ SCF CI method to study the excited states of acrolein and furan. (See also their earlier work (341).)

Del Bene and Jaffé (342) continued their series and computed the first ionization potential of azabenzenes with their modified CNDO method (see chapter 2, section 8). (Earlier, Bloor and Breen (343) applied the original CNDO parametrization to these molecules.) In a subsequent paper (344), they treated formaldehyde, formic acid, formamide, allene, butene, and diazomethane.

Giessner-Prettre and Pullman (345) studied the effect of σ and π singly excited configurations on electronic excited states using both the CNDO/2 and the INDO approximations. They applied these methods in the original parametrization of Pople et al. without fitting to experimental spectral data. Their target molecules were C_2H_4, H_2CO, HCOOH, and $HCONH_2$. Their CI matrices contained up to 80 singly excited configurations of a given symmetry. Their conclusions were that although these approximations do not give good values for the transition energies, they interpret well certain general properties of the spectra. These concern mainly the mixing of σ^* and π^* states and the existence of a large number of low-lying $\sigma \rightleftarrows \pi$ transitions.

Hunt and Goddard (346) computed the transition energies of the water molecule using improved virtual orbitals. Kirby and Miller (347) made a CNDO study of the geometry of ethylene in its first singlet excited state and confirmed its 90° twisted structure.

Tai and Allinger (348) applied the VESCF method to the electronic spectra of pyridine, pyrrole, and furan; Stals et al. (349) used the same method for unconjugated aromatic nitramines.

Watson et al. (350) presented an extensive work on the electronic transition of ethylene derivatives. No discussion of the detailed assignments will be given here but we should like to underscore the originality of this method. It uses the EH scheme with Mulliken-Wolfsberg-Helmholtz parametrization but includes into the AO basis C3s̲ and sometimes 3p̲ orbitals.

Salahub and Sandorfy (351) introduced Rydberg AO into the CNDO/2 method. They modified slightly Del Bene and Jaffé's scheme and introduced H2s̲, H2p̲, C3s̲ and C3p̲ orbitals into the basis. First-order configuration interaction was applied including the thirty configurations of lowest energy.

This Rydberg CNDO calculation showed that the ground and first excited singlet and triplet states of normal paraffins have practically no Rydberg character but the higher excited states contain various amounts of Rydberg admixtures.* In highly branched paraffins, the first singlet-singlet band corresponds to an excited state of high Rydberg character explaining the

*We do not use united atom Rydberg language here.

characteristic difference (202, 203) between the spectra of straight and highly branched paraffins. No such low-lying Rydberg excited state exists in the normal paraffins. The high density of electronic states for all but the smallest saturated hydrocarbons is substantiated by the calculations. The first singlet-singlet transition in ethane turns out to be polarized perpendicularly to the C−C line in agreement with experiment (352), and the first ionization is seen to leave the hole in an orbital mainly populated in the C−H bonds for methane, ethane, and propane, but in C−C bonds for the higher normal paraffins, and in mixed C−H and C−C populated orbitals for the branched paraffins.

This field is still in an exploratory stage. Future developments should consider open-shell SCF calculations for individual excited states (see Kaldor and Shavitt (353)) and systematic consideration of the correlation problem. In this respect the work done on atomic systems by Westhaus and Sinanoğlu (354, 355), Öksüz and Sinanoğlu (356−58) and Schaefer et al. (359) should be important for subsequent calculations on molecules.

6. Calculations on hydrogen bonded systems

Energies of hydrogen bonds and charge distribution in hydrogen bonds were, for a long time, treated as problems of intermolecular interactions. However, since the start of the epoch of extensive calculations on σ-electron systems in 1965, the way was cleared for the treatment of hydrogen bonded systems by normal quantum chemical methods. A hydrogen bonded system is considered as one molecule in these calculations. The works of Rein, Ladik, and Harris were mentioned on p. 45.

Hoffmann's extended Hückel scheme was tested in a few cases with moderate success. (See Murthy and Rao (360), Murthy, Davis, and Rao (361), and Rein, Clarke, and Harris (362), for example.)

The CNDO/2 method turned out to be much more successful; we can mention here only a few applications of this method. Pullman and Berthod (363) applied the method to formamide and its dimers. They obtained reasonable stabilization energies (5.8 kcal/mole per hydrogen bond) and charge distributions, and the cyclic dimer was seen to be the most stable among the possible dimers. Murthy and Rao (360) studied the water dimer. They obtained an enthalpy of formation somewhat higher than, and an equilibrium distance very near to the experimental value. Murthy, Davis, and Rao (361) examined the case of methanol and formic acid. Hoyland and Kier (364) applied the CNDO/2 method to a series of hydrogen bonded systems like the water dimer, the ammonia dimer, polymeric hydrogen fluorides, and others.

In all these cases the equilibrium distances are generally somewhat too small and hydrogen bond energies too large. The relative stabilities of possible conformations are usually predicted correctly. On the whole they predict a slight charge transfer from the proton donor to the proton acceptor molecule.

Schuster (365) made an extensive study of the interaction between carbonyl and hydroxyl groups using the CNDO/2 method on the model system of formaldehyde-water. He computed potential curves as functions of the O–H bond length, the O–O distance, and the angle between the functional groups, all in fair agreement with experimental data. He made the interesting observation that, as a result of hydrogen bonding, "all molecular orbitals originating from formaldehyde appear at slightly lower energies, while the water orbitals are destabilized to some extent." Schuster carried out similar work on formic acid and its dimer.

Bolander, Kassner, and Zung (366) made semiempirical calculations of the hydrogen bond energy for water clusters in the vapor phase. Ocvirk, Ažman, and Hadži (367) applied the CNDO/2 method to the formic acid monomer and dimer, trifluoroacetic acid, and the hydrogen bis (trifluoroacetate) ion, and Ažman, Koller, and Hadži to polywater (368). Haas and Feinberg (369) studied models for the theory of hydrogen bonding (linear H_3^-), Shaw (370) the LiH . . . Li^+ system.

Ab initio calculations on hydrogen bonded systems are mentioned on p. 104.

For a critical evaluation the reader is referred to Kollman and Allen (371).

7. Integral calculations

Several papers appeared in 1969 which give new methods for obtaining certain types of integrals needed in quantum-chemical calculations. Since these could be important in semiempirical work we give a partial list of these. Some works on ionization potentials and electron affinities on which parameters can be based will also be mentioned. In this respect we have to under-score the importance of the ionization potential data obtained by photoelectron spectroscopy initiated by D. W. Turner (see for example Al-Joboury and Turner (206)) which give invaluable information on the sequence and nature of molecular orbitals and whose full impact on quantum chemistry is yet to be felt. (See Molecular Photoelectron Spectroscopy, D. W. Turner, C. Baker, A. D. Baker, and C. R. Brundle, Wiley-Interscience, London and New York, 1970.)

Mehler and Ruedenberg (372). Two-center exchange integrals between Slater type AO. ("A novel general formula . . . which expresses two-center charge distributions over STO as expansions over terms (expressed in) elliptical coordinates. The formula is applied to derive compact expressions for the calculation of two-center exchange integrals arising from large basis sets.")

Flannery and Levy (373). Simple analytic expression for general two-center coulomb integrals. (It has a simple analytic dependence on nuclear separation.)

Newton (374). Projection of diatomic differential overlap (PDDO). A method for approximating one- and two-electron integrals over STO involving two-center charge distributions.

Newton et al. (375). Comparison of Gaussian expansion and PDDO methods using minimal STO basis sets.

Kay and Silverstone (376). Analytical evaluation of multi-center integrals of r_{12}^{-1} with STO. Four-center integrals by Fourier transform method.

Harris (377). Rapid evaluation of coulomb integrals. The properties of Fourier transforms and single-center expansions of STO are used to obtain formulas for general coulomb integrals.

Roberts and Warren (378). Justification of the Mataga-Nishimoto approximation (91). (See also Saturno (135) and Coulson (136).)

Tyutyulkov et al. (379). An approximation of the two-center coulomb integrals in semiempirical LCAO methods (a generalized Mataga-Nishimoto procedure).

Billingsley and Bloor (380). A limited expansion method for electron repulsion integrals.

Linderberg (381). The use of the contour integral method in MO theory. (Cf. Coulson (382).)

Whittington and Bersohn (383). A Monte Carlo method for quantum chemistry (numerical evaluation of integrals).

Weare et al. (384). Hulthen approximations to $1\underline{s}$ and $2\underline{p}$ orbitals of atoms (involving $O\underline{s}$ and $1\underline{p}$ orbitals).

Zollweg (385). Electron affinities of the heavy elements.

Levison and Perkins (386). Valence-state ionization potentials and nonempirical one-center electron repulsion integrals are evaluated for the separate $3\underline{s}$, $3\underline{p}$, and $3\underline{d}$ orbitals for second-row elements (for use as parameters in CNDO calculations).

8. Ab initio calculations (Gaussians)

Ab initio calculations are beyond our scope, but since there is an unmistakable trend to replace semiempirical methods by approximate ab initio calculations based in most cases on Gaussian atomic orbitals, the following list provides entries to the literature.

The floating spherical Gaussian orbitals (FSGO) model of molecular structure was initiated by Frost (387–89). Recently Rouse and Frost (390) modified the FSGO model to improve molecular energies and geometry by using a linear combination of two concentric spherical Gaussian orbitals (a double Gaussian). They approximated HF energies to 96% in the cases of H_2 and a number of hydrides and hydrocarbons.

Huzinaga and Sakai (391, 392) recommend the use of linear combinations of Gaussian-type orbitals (CGTO) instead of individual Gaussian-type orbitals, "as a unit of basis functions for large scale molecular calculations." They constructed such basis functions for the atoms from lithium through argon. Stewart (393) developed small Gaussian expansions for SCF AO for first-row atoms.

Neumann and Moskovitz (394, 395) used a contracted Gaussian basis for the study of one-electron properties of near-HF functions (application to acetaldehyde).

Petke et al. (396) reported Gaussian lobe function expansions for HF solutions for second row atoms. (See also Preuss (397).)

Whitten and Hackmeyer (398) carried out large scale configuration interaction calculations on formaldehyde using large Gaussian basis sets with a new method of generating configurations. Pedersen et al. (399) did this on excited states of nitrogen heterocyclic molecules. Petke and Whitten (400) examined orbital hybridization in polyatomic molecules.

Sachs, Geller, and Kaufman (401, 402) used Gaussian basis functions in their SCF MO investigation on the ground state of the CF_2 molecule and on the helium adduct of lithium hydride.

Veillard (403) also employed a Gaussian basis set for molecular wavefunctions containing second-row atoms. It consisted of 12 s-type and 9 p-type functions and it has been optimized. He reported energy values for this set and for different contractions of it.

Klessinger (404) introduced Gaussian expansions of minimal Slater-type orbital bases for calculations on molecular problems. Lathan et al. (405) made ab initio calculations using such a minimal basis set of Slater-type orbitals on dissociation energies of small hydrocarbons. Hehre et al. (406) gave a more general discussion of the method, and Ditchfield et al. (407) studied bond separation reactions using a contracted Gaussian basis.

André (408) published a SCF treatment for the electronic structure of polymers on a Gaussian basis, André et al. (409) on the barrier to internal rotation in allene, Letcher et al. (410) on ketene, with a uncontracted set of Gaussian 1s and 2p orbitals. Finally we mention a work by Switkes et al. (411) on polyatomic molecules using anisotropic basis sets of Slater-type orbitals (different exponents for $2p_x$, $2p_y$, and $2p_z$), and the ab initio calculations involving 3d orbitals in some sulfur compounds by Hillier and Saunders (412).

Recently several important papers were published on the application of ab initio methods to hydrogen bonded systems. We mention particularly the monumental attack of Clementi (413, 414) on the NH_3 + HCl system.

The water dimer attracted much attention. Morokuma and Pedersen (415) made LCAO MO SCF calculations using a medium sized Gaussian basis set (three s-type orbitals centered on the hydrogen atom and five s-type and three p-type orbitals for the oxygen atom). In a later paper, Morokuma and Winick (416) reexamined the water dimer using a minimal Slater basis set with exponents optimized for the water monomer. Large, contracted basis sets were used by Kollman and Allen (371), whose basis set consisted of 10 s and 5 p Gaussian functions for oxygen and 5 s functions for hydrogen. Diercksen's (417) basis set contained 11 s- and 7 p-type functions for oxygen and 6 s-type functions for hydrogen. The geometry corresponding to the lowest energy was linear. The energy of stabilization obtained was 5.3 kcal/mole in Kollman and Allen's calculations and 4.84 in Diercksen's.

(The experimental value is about 5.0 kcal/mole.) Kollman and Allen found, like Schuster (365) that all molecular orbitals on the proton acceptor decreased in energy and all those on the proton donor increased. They pointed out that this is generally so not only for hydrogen bonds but for all donor—acceptor complexes.

Del Bene and Pople (418) computed the association energies for small water polymers. They used a minimal basis set ($1s$, $2s$, and $2p$ for oxygen and $1s$ for hydrogen) where every Slater-type atomic orbital was replaced by a least-square-fitted combination of four Gaussian orbitals (419). Using sets of five or six Gaussian orbitals instead of four did not change the intermolecular energies appreciably. They found that chain-like polymers are favored and that the hydrogen bond energies deviate considerably from additivity. For higher polymers, however, cyclic structures associated with a chain are more stable.

Hankins, Moskowitz, and Stillinger (420) examined the non-additivity of hydrogen bond energy in water.

Dreyfus, Maigret, and Pullman (421) made ab initio calculations on the dimer of formamide. Their most interesting results are related to modifications of electronic distribution in the dimer with respect to the monomers. The nitrogen and oxygen atoms gain electrons, while the carbon atom and the hydrogen of the bridge lose electrons. There is a large σ gain and a smaller π loss on the nitrogen, and a smaller σ loss and a larger π gain on the oxygen. The overlap populations decrease in the NH and CO bonds while they increase in the CN. They conclude that ". . . exploration of density contours are of prime importance in the interpretation of fine structural effects together with the Mulliken populations."

These approximate ab initio SCF calculations using large Gaussian basis sets will certainly yield many interesting new results.

9. Localized orbitals

More and more quantum chemists are attracted by the descriptive value of the chemical formula. The year 1969 saw a few attempts to incorporate it into approximate wave-mechanical theories.

Hoyland continued his series on ab initio bond-orbital calculations (422, 423) and presented an improved procedure for saturated hydrocarbons (424). He describes them in terms of C1s core, C—H, and C—C bond orbitals assuming that after Löwdin orthogonalization they "are at least a description of the localized orbitals which would result from a transformation of the SCF MO obtained utilizing the same basis set." He first took all the C—H bond functions to be identical and transferable from molecule to molecule. The core and bond orbitals are written as linear combinations of a minimal set of carbon 1s, 2s, and 2p, and hydrogen 1s AO all of these expanded in terms of a basis set of spherical Gaussians (397). The total wavefunction is an antisymmetrized product of doubly occupied orthogonalized functions built up from these core and bond orbitals.

In his more recent work (424) Hoyland improved the basis set by using fully contracted nuclear-centered Gaussian sets. He found that the assumption of transferability of the C—H functions constitutes a deficiency of the method. Different functions should be used for primary, secondary, and tertiary C—H bonds.

He applied this method to barriers to internal rotation in ethane, propane, butane, and propylene and found considerable interaction between the two methyl groups in propane. He compared the energy values and populations to the results of SCF calculations. In another work (425), he computed the relative energies of some conformations of cyclopentane and cyclohexane.

Another bond-function analysis of rotational barriers in ethane came from Sovers et al. (426). The barrier was computed by integration of the Hellmann-Feynman forces along a path requiring only the force differences between the two conformations.

A frequent objection to molecular orbital methods is that the final wavefunction is not understandable in terms of chemical concepts. Baba et al. (427) described a procedure which makes it possible to translate the results of LCAO SCF PPP calculations into the language of a localized orbital, or other approximate models (locally excited states, charge transfer states, etc.).

Another localized-orbital scheme for σ bonds was proposed by Letcher and Dunning (428). The polyatomic molecule is described as a superposition of diatomic molecules of which a polyatomic wavefunction is synthesized. The diatomic angular momentum quantum numbers remain good quantum numbers and are used in the procedure.

Unland et al. (429) examined ground-state wavefunctions from the point of view of dipole moments and field gradients.

Rothenberg (430) applied the Edmiston-Ruedenberg localization criteria to ab initio SCF wavefunctions of methane, ethane, and methanol. He found the C–H orbitals practically identical in these molecules.

Polak (431) proposed a method for σ systems called SLO where the "strictly localized orbitals" are expressed in a minimum basis set of hybrid orbitals.

Peters (432) worked out a procedure in which the conventional HF method is slightly modified so that it yields localized molecular orbitals (LMO). Then perturbation theory is used. The final wavefunction is a single determinant of localized MO's. The whole work constitutes a major attempt to unify chemical valence theory and molecular quantum mechanics. So far it has been applied to methane.

Diner et al. (433) continued to develop their localized bond orbital theory, investigating its relation to the correlation problem. First they presented a perturbation calculation of the ground state energy, followed by an application to π-electron systems. Then Diner et al. (434) applied this method to σ systems and Jordan et al. (435) examined the stability of the perturbation energies with respect to bond hybridization and polarity. The chemical formula appears as the first approximation in the wave-mechanical treatment. The intention is then to go beyond SCF results without any variational procedure. First a set of reasonable bonding and antibonding orbitals is chosen, which are localized in the chemical bonds. The bonding orbitals are used to build a fully localized determinant which is the zero-order wavefunction. The antibonding orbitals are used to build the excited states. The SCF procedure is not employed; instead, the configurations obtained in this manner are used to set up a configuration-interaction matrix and the lowest eigenvalues are computed by a Rayleigh-Schrödinger perturbation expansion.

10. Valence-bond calculations

There has been renewed interest in recent years in the capabilities of the valence-bond method. This field probably deserves a separate review. Here we only mention some recent work related to our main subject matter.

Much of it goes back to the work of Porter and Karplus in 1964 (436). Karplus and Bersohn (437) studied the photodissociation of methane. Porter and Raff (438, 439) introduced configuration interaction in the simple valence-bond wavefunction and studied the potential energy surface in certain reactions involving σ electrons. Salomon (440) computed a semiempirical energy surface for H_3. Doggett (441) discussed atomic valence states in both valence-bond and molecular-orbital theory.

Shull (442) treated the calculation of matrix elements for valence-bond functions. He derived Pauling's formulas by a substitution process and generalized them. They do not depend upon the orthogonality of atomic orbitals nor upon the choice of the bond structures which may or may not be canonical.

References

1. C. Sandorfy and R. Daudel, Compt. Rend. **238**, 93 (1954).
2. C. Sandorfy, Can. J. Chem. **33**, 1337 (1955).
3. N. D. Sokolov, Rus. Chem. Rev. **34**, 960 (1967) (in English).
4. H. Yoshizumi, Trans. Faraday Soc. **53**, 125 (1957).
5. K. Fukui, H. Kato, and T. Yonezawa, Bull. Chem. Soc. Japan **33**, 1197, 1201 (1960).
6. K. Fukui, H. Kato, and T. Yonezawa, Bull. Chem. Soc. Japan **34**, 442, 1111 (1961).
7. G. Klopman, Helv. Chim. Acta **45**, 711 (1962).
8. G. Klopman, Helv. Chim. Acta **46**, 1967 (1963).
9. G. Klopman, Tetrahedron **19**, Suppl. 2, 111 (1963).
10. G. Klopman, Sur la structure électronique des molécules saturées, Cyanamid European Research Institute, Geneva, 1962.
11. G. W. Wheland, J. Am. Chem. Soc. **63**, 2025 (1941).
12. K. Fukui, H. Kato, T. Yonezawa, K. Morokuma, A. Imamura, and C. Nagata, Bull. Chem. Soc. Japan **35**, 38 (1962).
13. K. Fukui, in Modern Quantum Chemistry, Part I: Orbitals. Edited by O. Sinanoğlu. Academic Press, New York, 1965, pp. 49–84.
14. K. Fukui, in Molecular Orbitals in Chemistry, Physics and Biology. Edited by P. O. Löwdin and B. Pullman. Academic Press, New York, 1964, pp. 513–37.
15. G. Del Re, J. Chem. Soc., 4031 (1958).
16. G. Del Re, in Electronic Aspects of Biochemistry. Edited by B. Pullman. Academic Press, New York, 1964, pp. 221–35.
17. H. Cambron-Brüderlein and C. Sandorfy, Theoret. Chim. Acta **4**, 224 (1966).
18. R. S. Mulliken, J. Chim. Phys. **46**, 497, 675 (1949).
19. M. Wolfsberg and L. Helmholtz, J. Chim. Phys. **20**, 837 (1952).
20. G. Pilcher and H. A. Skinner, J. Inorg. Nucl. Chem. **24**, 937 (1962).
21. J. Hinze and H. H. Jaffé, J. Am. Chem. Soc. **84**, 540 (1962).
22. R. Hoffmann, J. Chem. Phys. **39**, 1397)1963).
23. A. D. Walsh, J. Chem. Soc., 2260 (1953).
24. R. Hoffmann, J. Chem. Phys. **40**, 2745 (1964).
25. R. Hoffmann, J. Chem. Phys. **40**, 2474 (1964).
26. R. Hoffmann, J. Chem. Phys. **40**, 2480 (1964).
27. R. Hoffmann, A. Imamura, and W. J. Hehre, J. Am. Chem. Soc. **90**, 1499 (1968).
28. W. Adam and A. Grimison, Tetrahedron **21**, 3417 (1965).
29. W. Adam and A. Grimison, Tetrahedron **22**, 835 (1966).
30. W. Adam, A. Grimison, and R. Hoffmann, J. Am. Chem. Soc. **91**, 2590 (1969).

31. J. Paldus and P. Hrabé, Theoret. Chim. Acta **11**, 401 (1968).
32. F. Jordan and B. Pullman, Theoret. Chim. Acta **9**, 242 (1968).
33. A. C. Hopkinson, R. A. McClelland, K. Yates, and I. G. Csizmadia, Theoret. Chim. Acta **13**, 65 (1969).
34. A. S. N. Murthy, R. E. Davis and C. N. R. Rao, Theoret. Chim. Acta **13**, 81 (1969).
35. L. C. Allen and J. D. Russell, J. Chem. Phys. **46**, 1029 (1967).
36. C. J. Ballhausen and H. B. Gray, Inorg. Chem. **1**, 111 (1962).
37. P. and R. Daudel, J. Phys. Radium **7**, 12 (1946).
38. A. Laforgue, J. Chim. Phys. **46**, 568 (1949).
39. G. W. Wheland and D. E. Mann, J. Chem. Phys. **17**, 264 (1949).
40. A. Streitwieser, J. Am. Chem. Soc. **82**, 4123 (1960).
41. A. Viste and H. B. Gray, Inorg. Chem. **3**, 1113 (1964).
42. G. Berthier, H. Lemaire, A. Rassat, and A. Veillard, Theoret. Chim Acta **3**, 213 (1965).
43. L. C. Cusachs and J. W. Reynolds, J. Chem. Phys. **43**, S 160 (1965).
44. R. Rein, N. Fukuda, H. Win, G. A. Clarke, and F. Harris, J. Chem. Phys. **45**, 4743 (1966).
45. E. Clementi and D. L. Raimondi, J. Chem. Phys. **38**, 2686 (1963).
46. L. C. Cusachs and B. B. Cusachs, J. Phys. Chem. **71**, 1060 (1967).
47. K. Ruedenberg, Rev. Mod. Phys. **34**, 326 (1962).
48. K. Ruedenberg and C. Edmiston, J. Phys. Chem. **68**, 1628 (1964).
49. L. C. Cusachs, B. L. Trus, D. G. Carroll, and S. P. McGlynn, Intern. J. Quantum Chem. **1 S**, 423 (1967).
50. J. C. Slater, Quantum Theory of Atomic Structure, Vol. 1, McGraw-Hill, New York, 1960, pp. 227–29.
51. K. Ruedenberg, J. Chem. Phys. **34**, 1892 (1961).
52. L. C. Cusachs, J. Chem. Phys. **43**, S 157 (1965).
53. D. G. Carroll, A. T. Armstrong, and S. P. McGlynn, J. Chem. Phys. **44**, 1865 (1966).
54. L. C. Cusachs, Intern. J. Quantum Chem., **1S**, 419 (1967).
55. B. J. Duke, Theoret. Chim. Acta **9**, 260 (1968).
56. M. Pollak and R. Rein, J. Chem. Phys. **47**, 2045 (1967).
57. R. S. Mulliken, J. Chem. Phys. **23**, 1833 (1955).
58. P. O. Löwdin, J. Chem. Phys. **18**, 365 (1950).
59. L. C. Cusachs and P. Politzer, Chem. Phys. Letters **1**, 529 (1968).
60. P. Politzer and L. C. Cusachs, Chem. Phys. Letters **2**, 1 (1968).
61. M. Roux, S. Besnainou and R. Daudel, J. Chim. Phys. **53**, 218, 939 (1956).
62. I. H. Hillier and J. F. Wyatt, Intern. J. Quantum Chem. **3**, 67 (1969).
63. A. Pullman, E. Kochanski, M. Gilbert, and A. Denis, Theoret. Chim. Acta **10**, 231 (1968).
64. A. Pullman, Intern. J. Quantum Chem. **2**, 187 (1968).
65. J. Ladik and G. Biczó, J. Chem. Phys. **42**, 1658 (1965); **49**, 1989 (1968).
66. I. Fischer-Hjalmars, B. Grabe, B. Roos, P. N. Skancke, and M. Sundbom, Intern. J. Quantum Chem. **1S**, 233 (1967).
67. W. Adam and A. Grimison, Theoret. Chim. Acta **7**, 342 (1967).

68. H. D. Bedon, S. M. Horner, and S. Y. Tyree, Inorg. Chem. 3, 647 (1964).
69. H. D. Bedon, W. E. Hatfield, S. M. Horner, and S. Y. Tyree, Inorg. Chem. 5, 743 (1965).
70. R. F. Fenske and C. C. Sweeney, Inorg. Chem. 3, 1105 (1964).
71. F. A. Cotton and T. E. Haas, Inorg. Chem. 3, 1004 (1964).
72. R. F. Fenske, Inorg. Chem. 4, 33 (1965).
73. G. Berthier, P. Millie, and A. Veillard, J. Chim. Phys. 62, 8, 20 (1965).
74. Chr. Klixbüll-Jörgensen, S. M. Horner, W. E. Hatfield, and S. Y. Tyree, Intern. J. Quantum Chem. 1, 191 (1967).
75. R. F. Fenske, K. G. Caulton, D. D. Radtke, and C. C. Sweeney, Inorg. Chem. 5, 951 (1966).
76. D. D. Radtke and R. F. Fenske, J. Am. Chem. Soc. 89, 2292 (1967).
77. M. D. Newton, F. P. Boer, and W. N. Lipscomb, J. Am. Chem. Soc. 88, 2353, 2367 (1966).
78. W. E. Palke and W. N. Lipscomb, J. Am. Chem. Soc. 88, 2384 (1966).
79. F. P. Boer, M. D. Newton, and W. N. Lipscomb, J. Am. Chem. Soc. 88, 2361 (1966).
80. R. D. Brown and R. D. Harcourt, Australian J. Chem. 16, 737 (1963).
81. J. A. Pople and D. P. Santry, Mol. Phys. 7, 269 (1963–64).
82. K. S. Pitzer and E. Catalano, J. Am. Chem. Soc. 78, 4844 (1956).
83. R. D. Brown and M. L. Heffernan, Trans. Faraday Soc. 54, 757 (1958).
84. R. D. Brown and M. L. Heffernan, Australian J. Chem. 12, 319, 330, 543, 554 (1959).
85. R. D. Brown and M. L. Heffernan, Australian J. Chem. 13, 38, 49 (1960).
86. R. Pariser and R. G. Parr, J. Chem. Phys. 21, 466, 767 (1953).
87. R. Pariser, J. Chem. Phys. 21, 568 (1953).
88. R. G. Parr, J. Chem. Phys. 20, 1499 (1952).
89. L. Paoloni, Nuovo Cimento 4, 410 (1956).
90. R. D. Brown and R. D. Harcourt, Australian J. Chem. 16, 737 (1963).
91. N. Mataga and K. Nishimoto, Z. Physik. Chem. 13, 140 (1957).
92. R. D. Brown and R. D. Harcourt, Australian J. Chem. 18, 1115 (1965).
93. R. D. Brown and J. B. Peel, Australian J. Chem. 21, 2589, 2605, 2617 (1968).
94. N. L. Allinger, J. C. Tai, and T. W. Stuart, Theoret. Chim. Acta 8, 101 (1967).
95. N. L. Allinger and J. C. Tai, J. Amer. Chem. Soc. 87, 1227 (1965).
96. R. McWeeny and T. E. Peacock, Proc. Phys. Soc. (London) A70, 41 (1957).
97. R. S. Mulliken, J. Phys. Chem. 56, 295 (1952).
98. J. A. Pople, D. P. Santry, and G. A. Segal, J. Chem. Phys. 43, S 129 (1965).
99. J. A. Pople and G. A. Segal, J. Chem. Phys. 43, S136 (1965).
100. J. A. Pople, Trans. Faraday Soc. 49, 1375 (1953).
101. D. B. Cook, P. C. Hollis, and R. McWeeny, Mol. Phys. 13, 553 (1967).
102. J. J. Kaufman, J. Chem. Phys. 43, S 152 (1965).

103. J. A. Pople and G. A. Segal, J. Chem. Phys. **44**, 3289 (1966).
104. J. A. Pople and M. Gordon, J. Am. Chem. Soc. **89**, 4253 (1967).
105. D. P. Santry and G. A. Segal, J. Chem. Phys. **47**, 158 (1967).
106. R. D. Brown and F. R. Burden, Theoret. Chim. Acta **12**, 95 (1968).
107. J. E. Bloor, B. R. Gilson, and F. P. Billingsley, Theoret. Chim. Acta **12**, 360 (1968).
108. J. A. Pople, D. L. Beveridge, and P. A. Dobosh, J. Chem. Phys. **47**, 2026 (1967).
109. R. N. Dixon, Mol. Phys. **12**, 83 (1967).
110. M. S. Gordon and J. A. Pople, J. Chem. Phys. **49**, 4643 (1968).
111. M. D. Newton, N. W. Ostlund, and J. A. Pople, J. Chem. Phys. **49**, 5192 (1968).
112. G. Klopman, J. Am. Chem. Soc. **86**, 1463 (1964).
113. C. E. Moore, Natl. Bur. Stand. Circ. 467, U.S. Govt. Printing Office, Washington, D.C., 1949.
114. H. O. Pritchard, Chem. Rev. **52**, 529 (1953).
115. B. Edlén, J. Chem. Phys. **33**, 98 (1960).
116. R. P. Iczkowski and J. L. Margrave, J. Am. Chem. Soc. **83**, 3547 (1961).
117. J. Hinze, M. A. Whitehead, and H. H. Jaffé, J. Am. Chem. Soc. **85**, 148 (1963).
118. G. Klopman, J. Am. Chem. Soc. **86**, 4550 (1964).
119. G. Klopman, J. Am. Chem. Soc. **87**, 3300 (1965).
120. R. T. Sanderson, Science **114**, 670 (1951).
121. R. Ferreira, Trans. Faraday Soc. **59**, 1064, 1075 (1963).
122. M. J. S. Dewar and G. Klopman, J. Am. Chem. Soc. **89**, 3089 (1967).
123. N. C. Baird and M. J. S. Dewar, J. Am. Chem. Soc. **89**, 3966 (1967).
124. N. C. Baird and M. J. S. Dewar, Theoret. Chim. Acta **9**, 1 (1967).
125. R. L. Flurry, Theoret. Chim. Acta **9**, 96 (1967).
126. M. J. S. Dewar, J. A. Hashmall, and C. G. Venier, J. Am. Chem. Soc. **90**, 1953 (1968).
127. L. Oleari, L. Di Scipio, and G. de Michaelis, Mol. Phys. **10**, 97 (1966).
128. J. M. Sichel and M. A. Whitehead, Theoret. Chim. Acta **7**, 32 (1967).
129. J. Hinze and H. H. Jaffé, J. Phys. Chem. **67**, 1501 (1963).
130. J. M. Sichel and M. A. Whitehead, Theoret. Chim. Acta **11**, 220 (1968).
131. N. Mataga, Bull. Chem. Soc. Japan **31**, 453 (1958).
132. K. Ohno, Theoret. Chim. Acta **2**, 219 (1964).
133. N. L. Baird, J. M. Sichel, and M. A. Whitehead, Theoret. Chim. Acta **11**, 38 (1968).
134. F. E. Harris, J. Chem. Phys. **48**, 4027 (1968).
135. A. F. Saturno, Theoret. Chim. Acta **11**, 365 (1968).
136. C. A. Coulson, Theoret. Chim. Acta **12**, 341 (1968).
137. R. G. Pearson and H. B. Gray, Inorg. Chem. **2**, 358 (1963).
138. R. Ferreira, J. Phys. Chem. **68**, 2240 (1964).
139. D. Peters, J. Chem. Soc. **A**, 644, 652, 656 (1966).

140. R. G. Pearson, J. Chem. Phys. **17**, 969 (1949).
141. J. R. Arnold, J. Chem. Phys. **24**, 181 (1956).
142. A. Veillard and G. Berthier, Theoret. Chim. Acta **4**, 347 (1966).
143. J. F. Labarre, M. Graffeuil, J.-P. Faucher, M. Pasdeloup, and J.-P. Laurent, Theoret. Chim. Acta **11**, 423 (1968).
144. B. Tinland, Theoret. Chim. Acta **11**, 452 (1968).
145. J. Kroner and H. Bock, Theoret. Chim. Acta **2**, 214 (1968).
146. H. A. Pohl, R. Rein, and K. Appel, J. Chem. Phys. **41**, 3385 (1964).
147. F. E. Harris and H. A. Pohl, J. Chem. Phys. **42**, 3648 (1965).
148. R. Rein and J. Ladik, J. Chem. Phys. **40**, 2466 (1964).
149. R. Rein and F. E. Harris, J. Chem. Phys. **41**, 3393 (1964).
150. R. Rein and F. E. Harris, J. Chem. Phys. **42**, 2177 (1965).
151. R. Rein and F. E. Harris, J. Chem. Phys. **43**, 4415 (1965); **45**, 1797 (1966).
152. R. Rein and F. E. Harris, Science **146**, 649 (1964).
153. R. Manne, Theoret. Chim. Acta **6**, 299, 312 (1966).
154. W. M. MacIntyre and P. O. Löwdin, Intern. J. Quant. Chem. **2 S**, 207 (1968).
155. R. Rein and S. Svetina, Intern. J. Quant. Chem. **1 S**, 171 (1967).
156. T. Yonezawa, K. Yamaguchi, and H. Kato, Bull. Chem. Soc. Japan **40**, 536 (1967).
157. H. Kato, H. Konishi, and T. Yonezawa, Bull. Chem. Soc. Japan **40**, 1017 (1967).
158. J. Hinze and H. H. Jaffé, J. Chem. Phys. **38**, 1834 (1963).
159. H. Kato, H. Konishi, H. Yamabe, and T. Yonezawa, Bull. Chem. Soc. Japan **40**, 2761 (1967).
160. J. A. Pople and R. K. Nesbet, J. Chem. Phys. **22**, 571 (1954).
161. T. Yonezawa, H. Nakatsuji, T. Kawamura, and H. Kato, Bull. Chem. Soc. Japan **40**, 2211 (1967).
162. T. Yonezawa, H. Kato, and H. Konishi, Bull. Chem. Soc. Japan **40**, 1071 (1967).
163. T. Yonezawa, H. Nakatsuji, and H. Kato, J. Am. Chem. Soc. **90**, 1239 (1968).
164. I. Fischer-Hjalmars, in Molecular Orbitals in Chemistry, Physics and Biology. Edited by P.-O. Löwdin and B. Pullman. Academic Press, New York, 1964, pp. 361–84.
165. P. M. Skancke, Arkiv F. Fysik **29**, 573 (1965).
166. P. N. Skancke, Arkiv F. Fysik **30**, 449 (1965).
167. J. C. Slater, Phys. Rev. **34**, 1293 (1929).
168. E. U. Condon and G. H. Shortley, The Theory of Atomic Spectra, Cambridge University Press, 1953, pp. 174–93.
169. W. Bingel, Z. Naturforsch. **9a**, 675 (1954).
170. W. E. Duncanson and C. A. Coulson, Proc. Roy. Soc. Edinburgh **A62**, 37 (1943).
171. E. Clementi, J. Chem. Phys. **38**, 2248 (1963).
172. W. Kolos and C. C. J. Roothaan, Rev. Mod. Phys. **32**, 205, 219 (1960).
173. M. P. Barnett, F. W. Birss, and C. A. Coulson, Mol. Phys. **1**, 44 (1958).

174. F. H. Field and J. L. Franklin, Electron Impact Phenomena, Academic Press, New York, 1957.
175. T. Koopmans, Physica I, 104 (1933).
176. M. Goeppert-Mayer and A. L. Sklar, J. Chem. Phys. 6, 645 (1938).
177. C. Fisk and S. Fraga, Can. J. Phys. 46, 1140, 2228 (1968).
178. D. B. Cook, P. C. Hollis, and R. McWeeny, Mol. Phys. 13, 553 (1967).
179. M. Klessinger and R. McWeeny, J. Chem. Phys. 42, 3343 (1965).
180. R. McWeeny and K. Ohno, Proc. Roy. Soc. London A255, 367 (1960).
181. R. McWeeny, Proc. Roy. Soc. London A253, 367 (1960).
182. R. McWeeny and G. Del Re, Theoret. Chim. Acta 10, 13 (1968).
183. G. Berthier, G. Del Re, and A. Veillard, Nuovo Cimento 44, 315 (1966).
184. G. Del Re and R. G. Parr, Rev. Mod. Phys. 35, 604 (1963).
185. M. Carpentieri, L. Porro, and G. Del Re, Intern. J. Quantum Chem. 2, 807 (1968).
186. A. Ciampi and L. Paoloni, Theoret. Chim. Acta 12, 229 (1968).
187. O. Sinanoğlu and B. Skutnik, Chem. Phys. Letters 1, 699 (1968).
188. C. Trindle and O. Sinanoğlu, J. Chem. Phys. 49, 65 (1968).
189. C. Trindle and O. Sinanoğlu, J. Am. Chem. Soc. 91, 853 (1969).
190. G. G. Hall, Proc. Roy. Soc. (London) A202, 336 (1959); A205, 541 (1951).
191. G. G. Hall and J. E. Lennard-Jones, Proc. Roy. Soc. (London) A202, 155 (1950); A205, 357 (1951).
192. J. E. Lennard-Jones and J. A. Pople, Proc. Roy. Soc. (London) A220, 446 (1950); A210, 190 (1951).
193. C. E. Edmiston and K. Ruedenberg, Rev. Mod. Phys. 34, 457 (1963).
194. C. E. Edmiston and K. Ruedenberg, J. Chem. Phys. 43, S97 (1965).
195. C. E. Edmiston and K. Ruedenberg, in Quantum Theory of Atoms, Molecules and the Solid State. Edited by P. O. Löwdin. Academic Press, New York, 1966, pp. 263–80.
196. O. Pamuk and O. Sinanoğlu (to be published).
197. Y. I'Haya, in Advances in Quantum Chemistry, Vol. 1, Academic Press, New York, 1964, pp. 203–40.
198. M. K. Orloff and O. Sinanoğlu, J. Chem. Phys. 43, 49 (1965).
199. S. Katagiri and C. Sandorfy, Theoret. Chim. Acta 4, 203 (1966).
200. C. L. Pekeris, Phys. Rev. 112, 1649 (1958).
201. R. L. Flurry, E. W. Stout, and J. J. Bell, Theoret. Chim. Acta 8, 203 (1967).
202. J. W. Raymonda and W. T. Simpson, J. Chem. Phys. 47, 430 (1967).
203. B. A. Lombos, P. Sauvageau, and C. Sandorfy, J. Mol. Spectry. 24, 253 (1967).
204. R. S. Mulliken, J. Chem. Phys. 3, 517 (1935).
205. R. S. Mulliken, J. Am. Chem. Soc. 86, 3183 (1964).
206. M. I. Al-Joboury and D. W. Turner, J. Chem. Soc. B, 373 (1967).
207. G. Bessis, private communication.
208. R. D. Brown and V. G. Krishna, J. Chem. Phys. 45, 1482 (1966).
209. A. Imamura, M. Kodama, Y. Tagashira, and C. Nagata, J. Theoret. Biol. 10, 356 (1966).

210. J. Del Bene and H. H. Jaffé, J. Chem. Phys. **48**, 1807 (1968).
211. M. El-Sayed, M. Kasha, and J. Tanaka, J. Chem. Phys. **34**, 334, (1961).
212. D. W. Turner, Tetrahedron Letters, No. 35, 3419 (1967).
213. J. Del Bene and H. H. Jaffé, J. Chem. Phys. **48**, 4050 (1968).
214. J. Del Bene and H. H. Jaffé, J. Chem. Phys. **49**, 1221 (1968).
215. H. W. Kroto and D. P. Santry, J. Chem. Phys. **47**, 792 (1967).
216. P. A. Clark and J. L. Ragle, J. Chem. Phys. **46**, 4235 (1967).
217. J. W. Moskowitz and M. C. Harrison, J. Chem. Phys. **42**, 1726 (1965).
218. J. M. Shulman and J. W. Moskowitz, J. Chem. Phys. **43**, 3287 (1965).
219. P. S. Song and T. A. Moore, Intern. J. Quantum Chem. **1**, 1699 (1967).
220. P. S. Song, Intern. J. Quantum Chem. **2**, 281, 297 (1968).
221. N. L. Allinger, C. Gilardeau and L. W. Chow, Tetrahedron **24**, 2401 (1968).
222. A. Denis and J.-P. Malrieu, Theoret. Chim. Acta **12**, 66 (1968).
223. D. R. Salahub and C. Sandorfy (to be published).
224. H. Kato, H. Konishi, and T. Yonezawa, Bull. Chem. Soc. Japan **39**, 2774 (1966).
225. B. Skutnik, I. Öksüz, and O. Sinanoğlu, Intern. J. Quantum Chem. **2 S**, 1 (1968).
226. H. Hosoya, J. Chem. Phys. **48**, 1380 (1968).
227. R. S. Berry, J. Chem. Phys. **38**, 1934 (1963).
228. M. B. Robin, R. R. Hart, and N. A. Kuebler, J. Chem. Phys. **44**, 1803 (1966).
229. F. Gallais and D. Voigt, Bull. Soc. Chim. France 70, (1960).
230. P. Pascal, Ann. Chim. Phys. **19**, 70 (1910); P. Pascal, A. Pacault, and J. Hoarau, Compt. Rend. **233**, 1078 (1951).
231. R. Daudel, Compt. Rend. **237**, 601 (1953); R. Daudel, S. Odiot and H. Brion, J. Chim. Phys. **51**, 74 (1954); H. Brion, R. Daudel, and S. Odiot, J. Chim. Phys. **51**, 358 (1954); S. Odiot and R. Daudel, J. Chim. Phys. **51**, 361 (1954); R. Daudel, H. Brion, and S. Odiot, J. Chem. Phys. **23**, 2080 (1955).
232. R. Daudel, Les Fondements de la Chimie Théorique, Gauthier Villars, Paris, 1956 (or Fundamentals of Theoretical Chemistry, Pergamon Press, 1968); R. Daudel, Advan. Quantum Chem. **1**, 115 (1964).
233. R. Daudel, H. Brion, and S. Odiot, J. Chem. Phys. **23**, 2080 (1955).
234. S. Odiot, Cahiers Phys. **81**, 1 (1957); **82**, 23 (1957).
235. S. Odiot and R. Daudel, Compt. Rend. **238**, 1384 (1954).
236. M. Klessinger and R. McWeeny, J. Chem. Phys. **42**, 3343 (1965). See also: R. McWeeny, Proc. Roy. Soc. (London) **A253**, 242 (1959); **A259**, 554 (1961); Rev. Mod. Phys. **32**, 335 (1960); E. Kapuy, Acta Phys. Acad. Sci. Hung. **13**, 345 (1961); **15**, 177 (1962); **15**, 341 (1962); Physics Letters **1**, 205 (1962).
237. R. Daudel, F. Gallais, and P. Smet, Intern. J. Quantum Chem. **1**, 873 (1967).
238. V. A. Fock, Dokl. Akad. Nauk. SSSR **73**, 735 (1950).

239. A. C. Hurley, J. Lennard-Jones, and J. A. Pople, Proc. Roy. Soc. (London) **A220**, 446 (1953).
240. E. Kapuy, Acta Phys. Hung. **9**, 237 (1958).
241. E. Kroner, Z. Naturforsch. **15a**, 260 (1960); F. Bopp, Z. Physik **156**, 348 (1959).
242. R. McWeeny and B. T. Sutcliffe, Proc. Roy. Soc. (London) **A273**, 103 (1963).
243. E. V. Ludena and V. Amzel (personal communication).
244. T. Arai, J. Chem. Phys. **33**, 95 (1960).
245. P. O. Löwdin, J. Chem. Phys. **35**, 78 (1961).
246. P. Durand and R. Daudel, Cahiers de Physique **166**, 225 (1964).
247. K. Miller and K. Ruedenberg, communication to Sanibel Meeting (1965).
248. J. M. Leclercq, Thèse de 3ème cycle, Sorbonnne (1966).
249. R. Daudel and A. Veillard, in La Nature et les Propriétés des Liaisons de Coordination, CNRS Editions, Paris 1970.
250. P. Millié and G. Berthier, in La Nature et les Propriétés des Liaisons de Coordination, CNRS Editions, Paris 1970; M. Bigorgne (ibid.).
251. G. G. Hall, Proc. Phys. Soc. **A205**, 541 (1951).
252. D. M. Silver, J. Chem. Phys. **50**, 5108 (1969).
253. E. L. Mehler, K. Ruedenberg, and D. M. Silver, J. Chem. Phys. **52**, 1174, 1181 (1970).
254. D. M. Silver, K. Ruedenberg, and E. L. Mehler, J. Chem. Phys. **52**, 1206 (1970).
255. A. Ciampi and L. Paoloni, Theoret. Chim. Acta **12**, 229 (1968).
256. A. Veillard and R. Daudel, in La Nature et les Propriétés des Liaisons de Coordination, CNRS Editions, Paris 1970.
257. R. Ahlrichs and W. Kutzelnigg, J. Chem. Phys. **48**, 1819 (1968); Theoret. Chim. Acta **10**, 377 (1968); Chem. Phys. Letters **1**, 651 (1968).
258. E. A. Scarzafava, Ph.D. Thesis, Indiana University (1969).
259. R. Daudel and P. Durand, Mod. Quantum Chem. **2**, 75 (1965).
260. S. Bratoz, Compt. Rend. **256**, 5298 (1963). See also S. Bratoz and Ph. Durand, J. Chem. Phys. **8**, 2670 (1965).
261. G. Bessis, C. Murez, and S. Bratoz, Intern. J. Quantum Chem. **1**, 327 (1967).
262. G. Bessis, Ph. Espagnet, and S. Bratoz, Intern. J. Quantum Chem. **3**, 205 (1969).
263. R. D. Brown, J. Chem. Soc. 2615 (1953).
264. R. Daudel (unpublished theory).
265. S. Diner, J. P. Malrieu, and P. Claverie, Theoret. Chim. Acta, **13**, 1 (1969); S. Diner, J. P. Malrieu, F. Jordan, and M. Gilbert, Theoret. Chim. Acta **15**, 100 (1969).
266. J. Lennard-Jones and G. G. Hall, Trans. Faraday Soc. **48**, 581 (1952).
267. R. Thompson, Conference on Applied Mass Spectroscopy, Institute of Petroleum, London, 1953, p. 185.

268. N. D. Coggeshall, J. Chem. Phys. **30**, 595 (1959).
269. J. C. Lorquet, Mol. Phys. **9**, 101 (1965).
270. A. Julg, J. Chim. Phys. **53**, 548 (1956).
271. F. E. Harris, J. Chem. Phys. **48**, 4027 (1968).
272. K. Jug, J. Chem. Phys. **51**, 2779 (1969).
273. K. Jug, Theoret. Chim. Acta **16**, 95 (1970).
274. J. H. Corrington and L. C. Cusachs, Intern. J. Quantum Chem. **3s**, 207 (1969).
275. D. J. Miller and L. C. Cusachs, Chem. Phys. Letters **3**, 501 (1969).
276. R. Ferreira and J. K. Bates, Theoret. Chim. Acta **16**, 111 (1970).
277. J. J. Kaufman and J. J. Harkins, J. Chem. Phys. **50**, 771 (1969).
278. D. B. Chesnut and R. W. Moseley, Theoret. Chim. Acta **13**, 230 (1969).
279. W. A. Yeranos, Theoret. Chim. Acta **13**, 246 (1969).
280. W. H. de Jeu and G. P. Benader, Theoret. Chim. Acta **13**, 349 (1969).
281. S. F. A. Kettle and V. Tomlinson, Theoret. Chim. Acta **14**, 175 (1969).
282. L. B. Kier and J. M. George, Theoret. Chim. Acta **14**, 258 (1969).
283. A. Rossi, C. W. David, and R. Schor, Theoret. Chim. Acta **14**, 429 (1969).
284. B. J. Mc Aloon and B. C. Webster, Theoret. Chim. Acta **15**, 385 (1969).
285. J. F. Olsen and L. Burnelle, J. Am. Chem. Soc. **91**, 7286 (1969).
286. D. B. Boyd, Theoret. Chim. Acta **14**, 402 (1969).
287. D. B. Cook and P. Palmieri, Mol. Phys. **17**, 271 (1969).
288. K. R. Roby and O. Sinanoğlu, Intern. J. Quantum Chem. **3s**, 223 (1969).
289. R. D. Brown and K. R. Roby, On the Foundations of Approximate Molecular Orbital Theory (to be published).

290. G. Burns, J. Chem. Phys. **41**, 1521 (1964).
291. C. Hollister and O. Sinanoğlu, J. Am. Chem. Soc. **88**, 13 (1966).
292. P. Westhaus and O. Sinanoğlu, Intern. J. Quantum Chem. **3s**, 391 (1970).
293. K. Jug, Intern. J. Quantum Chem. **3s**, 241 (1969).
294. S. Fischer, Intern. J. Quantum Chem. **3s**, 651 (1970).
295. A. Rastelli and G. Del Re, Intern. J. Quantum Chem. **3**, 553 (1969).
296. G. Del Re, Intern. J. Quantum Chem. **1**, 293 (1967).
297. M. G. Hegyi, M. Mezei, and T. Szondy, Theoret. Chim. Acta **15**, 273 (1969).
298. J. Paldus and J. Čížek, Chem. Phys. Letters **3**, 1 (1969).
299. P. Gombás, Pseudopotentiale, Springer, Vienna, 1967.
300. P. Gombás, Theoret. Chim. Acta **5**, 112 (1966).
301. P. Gombás, Theoret. Chim. Acta **11**, 210 (1968).
302. P. Gombás and T. Szondy, Acta Phys. Acad. Sci. Hung. **25 (4)**, 345 (1968).
303. P. Gombás and T. Szondy (to be published).
304. W. Kutzelnigg, Chem. Phys. Letters **4**, 435 (1969).

118 References

305. G. McGinn, J. Chem. Phys. **50**, 1404 (1969).
306. G. McGinn, J. Chem. Phys. **51**, 5090 (1969).
307. K. B. Wiberg, J. Am. Chem. Soc. **90**, 59 (1968).
308. D. H. Lo and M. A. Whitehead, Can. J. Chem. **46**, 2027, 2041 (1968).
309. D. H. Lo and M. A. Whitehead, J. Chem. Soc. A 1513 (1969).
310. D. H. Lo and M. A. Whitehead, Tetrahedron **25**, 2615 (1969).
311. R. J. Boyd and M. A. Whitehead, J. Chem. Soc. A 2598 (1969).
312. H. Fischer and H. Kollmar, Theoret. Chim. Acta **13**, 213 (1969).
313. J. Linderberg, Chem. Phys. Letters **1**, 39 (1967).
314. D. W. Davies, Mol. Phys. **17**, 473 (1969).
315. K. Jug, Theoret. Chim. Acta **14**, 91 (1969).
316. M. J. S. Dewar and S. D. Worley, J. Chem. Phys. **50**, 654 (1969).
317. M. J. S. Dewar and S. D. Worley, J. Chem. Phys. **51**, 263 (1969).
318. N. C. Baird and M. J. S. Dewar, J. Chem. Phys. **50**, 1262 (1969).
319. N. C. Baird and M. J. S. Dewar, J. Am. Chem. Soc. **91**, 352 (1969).
320. N. C. Baird, M. J. S. Dewar, and R. Sustmann, J. Chem. Phys. **50**, 1275 (1969).
321. H. Betsuyaku, J. Chem. Phys. **50**, 3118 (1969).
322. G. A. Segal, R. Bruns, and W. B. Person, J. Chem. Phys. **50**, 3811 (1969).
323. G. W. King, D. P. Santry, and C. H. Warren, J. Chem. Phys. **50**, 4565 (1969).
324. D. T. Clark, Theoret. Chim. Acta **10**, 111 (1968).
325. D. T. Clark, Theoret. Chim. Acta **15**, 225 (1969).
326. T. Yonezawa, H. Katô, and H. Katô, Theoret. Chim. Acta **13**, 125 (1969).
327. A. L. Becker and J. P. Dahl, Theoret. Chim. Acta **14**, 26 (1969).
328. R. Carbo, M. S. Giambiagi, and M. Giambiagi, Theoret. Chim. Acta **14**, 147 (1969).
329. M. V. Basilevsky and I. E. Chlenov, Theoret. Chim. Acta **15**, 174 (1969).
330. B. Pullman and H. Berthod, Theoret. Chim. Acta **15**, 205 (1969).
331. G. H. Cheesman, A. J. T. Finney, and I. K. Snook, Theoret. Chim. Acta **16**, 33 (1970).
332. R. Tinland, Mol. Phys. **16**, 413 (1969).
333. A. Gamba, G. Morosi, and M. Simonetta, Chem. Phys. Letters **3**, 20 (1969).
334. J. L. Gole, Chem. Phys. Letters **3**, 577 (1969).
335. S. Ehrenson, Chem. Phys. Letters **3**, 585 (1969).
336. G. R. Branton, D. C. Frost, F. G. Herring, C. A. McDowell, and I. A. Stenhouse, Chem. Phys. Letters **3**, 581 (1969).
337. L. Asbrink, O. Edqvist, E. Lindholm, and L. E. Selin, Chem. Phys. Letters **5**, 192 (1970).
338. R. M. Canadine and I. H. Hillier, J. Chem. Phys. **50**, 2984 (1969).
339. A. Pullman, Intern. J. Quantum Chem. **2s**, 187 (1968).
340. M. Jungen and H. Labhart, Theoret. Chim. Acta **9**, 345 (1968).
341. M. Jungen, H. Labhart, and G. Wagnière, Theoret. Chim. Acta **4**, 305 (1966).

342. J. Del Bene and H. H. Jaffé, J. Chem. Phys. **50**, 563 (1969).
343. J. E. Bloor and D. L. Breen, J. Am. Chem. Soc. **89**, 6835 (1967).
344. J. Del Bene and H. H. Jaffé, J. Chem. Phys. **50**, 1126 (1969).
345. C. Giessner-Prettre and A. Pullman, Theoret. Chim. Acta **13**, 265 (1969).
346. W. J. Hunt and W. A. Goddard, Chem. Phys. Letters **3**, 414 (1969).
347. G. H. Kirby and K. Miller, Chem. Phys. Letters **3**, 643 (1969).
348. J. C. Tai and N. L. Allinger, Theoret. Chim. Acta **15**, 133 (1969).
349. J. Stals, C. G. Barraclough, and A. S. Buchanan, Trans. Faraday Soc. **65**, 904 (1969).
350. F. H. Watson, A. T. Armstrong, and S. P. McGlynn, Theoret. Chim. Acta **16**, 75 (1970).
351. D. R. Salahub and C. Sandorfy (to be published).
352. E. F. Pearson and K. K. Innes, J. Mol. Spectry. **30**, 232 (1969).
353. U. Kaldor and I. Shavitt, J. Chem. Phys. **48**, 191 (1968).
354. P. Westhaus and O. Sinanoğlu, Astrophys. J. **157**, 997 (1969).
355. P. Westhaus and O. Sinanoğlu, Phys. Rev. **183**, 56 (1969).
356. I. Öksüz and O. Sinanoğlu, Phys. Rev. **181**, 42 (1969).
357. I. Öksüz and O. Sinanoğlu, Phys. Rev. **181**, 54 (1969).
358. O. Sinanoğlu, Atomic Physics, Plenum Press, 1969, p. 131.
359. H. F. Schaefer, R. A. Klemm, and F. E. Harris, J. Chem. Phys. **51**, 4643 (1969).
360. A. S. N. Murthy and C. N. R. Rao, Chem. Phys. Letters **2**, 123 (1968).
361. A. S. N. Murthy, R. E. Davis, and C. N. R. Rao, Theoret. Chim. Acta **13**, 81 (1969).
362. R. Rein, G. A. Clarke, and F. E. Harris, J. Mol. Structure **2**, 103 (1968).
363. A. Pullman and H. Berthod, Theoret. Chim. Acta **10**, 461 (1968).
364. J. R. Hoyland and L. B. Kier, Theoret. Chim. Acta **15**, 1 (1969).
365. P. Schuster, Int. J. Quant. Chem. **3**, 851 (1969).
366. R. W. Bolander, J. L. Kassner, and J. T. Zung, J. Chem. Phys. **50**, 4402 (1969).
367. A. Ocvirk, A. Ažman, and D. Hadži, Theoret. Chim. Acta **10**, 187 (1968).
368. A. Ažman, J. Koller, and D. Hadži, Chem. Phys. Letters **5**, 157 (1970).
369. T. E. Haas and M. J. Feinberg, Theoret. Chim. Acta **10**, 189 (1969).
370. G. Shaw, Int. J. Quant. Chem. **3**, 219 (1969).
371. P. A. Kollman and L. C. Allen, J. Chem. Phys. **51**, 3286 (1969).
372. E. L. Mehler and K. Ruedenberg, J. Chem. Phys. **50**, 2575 (1969).
373. M. R. Flannery and H. Levy, J. Chem. Phys. **50**, 2938 (1969).
374. M. D. Newton, J. Chem. Phys. **51**, 3917 (1969).
375. M. D. Newton, W. A. Lathan, W. J. Hehre, and J. A. Pople, J. Chem. Phys. **51**, 3927 (1969).
376. K. G. Kay and H. J. Silverstone, J. Chem. Phys. **51**, 4287 (1969).
377. F. E. Harris, J. Chem. Phys. **51**, 4770 (1969).
378. G. Roberts and K. D. Warren, Theoret. Chim. Acta **13**, 353 (1969).

379. N. Tyutyulkov, A. Gochev, and F. Fratev, Chem. Phys. Letters **4**, 9 (1969).
380. F. P. Billingsley and J. E. Bloor, Chem. Phys. Letters **4**, 48 (1969).
381. J. Linderberg, Chem. Phys. Letters **5**, 134 (1970).
382. C. A. Coulson, Proc. Cambridge Phil. Soc. **36**, 21 (1940).
383. S. G. Whittington and M. Bersohn, Mol. Phys. **17**, 627 (1969).
384. J. H. Weare, T. A. Weber, and R. G. Parr, J. Chem. Phys. **50**, 4393 (1969).
385. R. J. Zollweg, J. Chem. Phys. **50**, 4251 (1969).
386. K. A. Levison and P. G. Perkins, Theoret. Chim. Acta **14**, 206 (1969).
387. A. A. Frost, J. Chem. Phys. **47**, 3707, 3714 (1967).
388. A. A. Frost, J. Phys. Chem. **72**, 1289 (1968).
389. A. A. Frost and R. A. Rouse, J. Am. Chem. Soc. **90**, 1965 (1968).
390. R. A. Rouse and A. A. Frost, J. Chem. Phys. **50**, 1705 (1969).
391. S. Huzinaga, J. Chem. Phys. **42**, 1293 (1965).
392. S. Huzinaga and Y. Sakai, J. Chem. Phys. **50**, 2465 (1969).
393. R. F. Stewart, J. Chem. Phys. **50**, 2465 (1969).
394. D. B. Neumann and J. W. Moskowitz, J. Chem. Phys. **49**, 2056 (1968).
395. D. B. Neumann and J. W. Moskowitz, J. Chem. Phys. **50**, 2216 (1969).
396. J. D. Petke, J. L. Whitten, and A. W. Douglas, J. Chem. Phys. **51**, 256 (1969).
397. H. Preuss, Z. Naturforsch. **11a**, 823 (1956).
398. J. L. Whitten and M. Hackmeyer, J. Chem. Phys. **51**, 5584 (1969).
399. L. Pedersen, D. G. Whitten, and M. T. McCall, Chem. Phys. Letters **3**, 569 (1969).
400. J. D. Petke and J. L. Whitten, J. Chem. Phys. **51**, 3166 (1969).
401. L. M. Sachs, M. Geller, and J. J. Kaufman, J. Chem. Phys. **51**, 2771, (1969).
402. J. J. Kaufman and L. M. Sachs, J. Chem. Phys. **51**, 2992 (1969).
403. A. Veillard, Theoret. Chim. Acta **12**, 405 (1968).
404. M. Klessinger, Theoret. Chim. Acta **15**, 353 (1969).
405. W. A. Lathan, W. J. Hehre, and J. A. Pople, Chem. Phys. Letters **3**, 579 (1969).
406. W. J. Hehre, R. F. Stewart, and J. A. Pople, J. Chem. Phys. **51**, 2657 (1969).
407. R. Ditchfield, W. J. Hehre, J. A. Pople, and L. Radom, Chem. Phys. Letters **5**, 13 (1970).
408. J. M. André, J. Chem. Phys. **50**, 1536 (1969).
409. J. M. André, M. C. André, and G. Leroy, Chem. Phys. Letters **3**, 695 (1969).
410. J. H. Letcher, M. L. Unland, and J. R. Van Wazer, J. Chem. Phys. **50**, 2185 (1969).
411. E. Switkes, R. M. Stevens, and W. N. Lipscomb, J. Chem. Phys. **51**, 5229 (1969).
412. I. H. Hillier and V. R. Saunders, Chem. Phys. Letters **4**, 163 (1969).
413. E. Clementi, J. Chem. Phys. **46**, 3851 (1967); **47**, 2323 (1967).
414. E. Clementi and J. N. Gayles, J. Chem. Phys. **47**, 3837 (1967).
415. K. Morokuma and L. Pedersen, J. Chem. Phys. **48**, 3275 (1968).

416. K. Morokuma and J. R. Winick, J. Chem. Phys. **52**, 1301 (1970).
417. G. H. F. Diercksen, Chem. Phys. Letters **4**, 373 (1969).
418. J. Del Bene and J. A. Pople, Chem. Phys. Letters **4**, 426 (1969).
419. W. J. Hehre, R. F. Stewart, and J. A. Pople, J. Chem. Phys. **51**, 2657 (1969).
420. D. Hankins, J. W. Moskowitz, and F. H. Stillinger, Chem. Phys. Letters **4**, 527 (1969).
421. M. Dreyfus, B. Maigret, and A. Pullman, Theoret. Chim. Acta **17**, 109 (1970).
422. J. R. Hoyland, J. Am. Chem. Soc. **90**, 2227 (1968).
423. J. R. Hoyland, J. Chem. Phys. **49**, 1908, 2563 (1968).
424. J. R. Hoyland, J. Chem. Phys. **50**, 473 (1969).
425. J. R. Hoyland, J. Chem. Phys. **50**, 2775 (1969).
426. O. J. Sovers, C. W. Kern, R. M. Pitzer, and M. Karplus, J. Chem. Phys. **49**, 2592 (1968).
427. H. Baba, S. Suzuki, and T. Takemura, J. Chem. Phys. **50**, 2078 (1969).
428. J. H. Letcher and T. H. Dunning, J. Chem. Phys. **48**, 4538 (1968).
429. M. L. Unland, T. H. Dunning, and J. R. Van Wazer, J. Chem. Phys. **50**, 3208, 3214 (1969).
430. S. Rothenberg, J. Chem. Phys. **51**, 3389 (1969).
431. R. Polak, Theoret. Chim. Acta **14**, 163 (1969).
432. D. Peters, J. Chem. Phys. **50**, 1559, 1566 (1969).
433. S. Diner, J. P. Malrieu, and P. Claverie, Theoret. Chim. Acta **13**, 1, 18 (1969).
434. S. Diner, J. P. Malrieu, F. Jordan, and M. Gilbert, Theoret. Chim. Acta **15**, 100 (1969).
435. F. Jordan, M. Gilbert, J. P. Malrieu, and U. Pincelli, Theoret. Chim. Acta **15**, 211 (1969).
436. R. N. Porter and S. Karplus, J. Chem. Phys. **40**, 1105 (1964).
437. S. Karplus and R. Bersohn, J. Chem. Phys. **51**, 2040 (1969).
438. R. N. Porter and L. M. Raff, J. Chem. Phys. **50**, 5216 (1969).
439. L. M. Raff and R. N. Porter, J. Chem. Phys. **51**, 4701 (1969).
440. M. Salomon, J. Chem. Phys. **51**, 2406 (1969).
441. G. Doggett, Theoret. Chim. Acta **15**, 344 (1969).
442. H Shull, Intern. J. Quantum Chem. **3**, 523 (1969).

Ab initio calculations, 103
AB$_2$ type molecules, 33
A$_2$Y$_4$ type molecules, 24
Acetylene, 52, 53, 77
Acrolein, 46, 98
Activation energies, 4
Additivity, of properties of
  paraffins, 5, 22
Adenozine triphosphate, 93
Alkyl cations, 46
Allene, 99, 104
Alternation of bond lengths, 33
Aminoacids, 93
Ammonia, 25, 31, 34, 52, 98
Ammoniated electron, 93
Angular momentum, 106
Antibonding orbitals, delocaliza-
  tion into, 24
Azabenzenes, 98
Azines, 12

β, proportionality to S,
  criticism of, 54
Barriers to internal rotation,
  10, 106
Basis set, choice of, 55
Bending force constants, 34
Benzene, 61, 62, 63, 64, 98
Benzynes, 12
Beryllium, 76, 79
Biological molecules, 16, 45, 98
Biorbitals, 76
Biphenyl, 43
Bond dissociation energies, 3,
  93
Boron compounds, 12, 19, 43, 82,
  93, 98
Butadiene, 8, 46
Butane, 7
Butene, 99

"C" approximation, 1
Carbonium ions, 12, 41, 46, 97
Carbon monoxide, 34
Carbon trioxide radical, 93
Carbonyl compounds, 19
Charges: bond, 8; iteration to,
  16, 34, 42, 91, 96; most prob-
  able distribution of, 82; net
  atom, 8; orbital, 3, 8; overlap,
  16; variation of, 16
CNDO/2, 33, 34, 55
Complete neglect of differential
  overlap, 26, 27, 96; criticism
  of, 30
Configuration interaction, 22
Conformations: of cycloparaffins,
  10; of nucleosides, 12
Conjugated molecules, 8; and loge
  functions, 76; order of orbitals
  in, 19; sigma bonds in, 8; small
  ring, 41
Coordination complexes, 17
Correlation. See electron correlation
Cusachs' resonance integral, 14,
  15
Cyclobutadiene, 63
Cyclopentadiene, 63
Cyclopropane, 58, 59, 98

Delocalization, 5; into antibonding orbitals, 24; of bonding
  electrons, 5, 20; causes of, 19;
  geminal, 22; of lone pairs, 19,
  24, 63; in paraffins, 5; vicinal,
  22
Density contours, 105
Desoxyribonucleic acid, 16
Diatomic hydrides, for calibration
  of bonding parameters, 41